ENGLISH
NOW AND THEN

By the same author

Leaving: an anthology for the last year at school, Hutchinson, 1965

Fourteen: Autobiography of an Age-Group, Cambridge, 1965

ENGLISH NOW AND THEN

BY D. C. MEASHAM

CAMBRIDGE
AT THE UNIVERSITY PRESS
1965

PUBLISHED BY
THE SYNDICS OF THE CAMBRIDGE UNIVERSITY PRESS
Bentley House, 200 Euston Road, London, N.W. 1
American Branch: 32 East 57th Street, New York, N.Y. 10022
West African Office: P.O. Box 33, Ibadan, Nigeria

Printed in Great Britain at the University Printing House, Cambridge (Brooke Crutchley, University Printer)

LIBRARY OF CONGRESS CATALOGUE CARD NUMBER: 65–19154

Contents

Foreword

I have heard it said that English as it is commonly taught is like owning a car which one spends one's time greasing and cleaning, instead of riding around in it. This present book has its fair share of 'maintenance work'—but it does not go in for trial and error tinkering, and does, I hope, put the vehicle to good use.

Some oral questions are included in the text, but I have not tried to do what the teacher in the classroom can do better. I have not, for instance, put in lengthy asides to explain 'difficult' words when they were the only ones which could be used naturally in that particular context.

The work for each passage is set out under three separate heads. Section A can be used for oral work, but was intended (in most cases) to be done in writing; all the necessary information is in the passage. Section B requires independent thought and, often, reference books. Section C suggests some side issues which can be explored in detail. A main aim has been to link language work and library research. I have found when doing the course in school that a satisfying summary results from all the work being done in a special 'language' exercise book; the continuous argument of the course would be lost if the record of work were to be done in rough, or interrupted by other aspects of English.

The book deals with the origins of language, its background, structure and effect, and introduces some main concepts in modern linguistics. Though the presentation of the work is, of course, my own, this is an eclectic book—and probably the better for that. The sources that I can think of are acknowledged either in the text itself or in the list of recommended reading.

The book represents a year's work and much of the material might be used as a 'shot in the arm' for intelligent children at any stage of their secondary school career. However, it is perhaps particularly suitable for the final 'statutory' year of school, or for the first year of a sixth form. It might also be a preparation for a rational 'O' level English Language paper, or a rational 'Use of English' paper.

I wish to thank Mr A. Griffiths, now of the Leeds Institute of Education, for his help and encouragement when I worked with him. My thanks go also to Mr V. Callaghan, of Netteswell Comprehensive School, and to Miss K. M. Quinn, Vice-Principal of Matlock College, who read the proofs and made valuable suggestions. Finally, I must acknowledge my debt to the editorial staff of the Cambridge University Press.

D. C. M.

Matlock College of Education,
May 1965

1

Open to Dispute from the Beginning

There is no general agreement as to how language began. About all that can be said without fear of contradiction is that a word is a *sound* or group of sounds when spoken, and a convenient way of representing a sound or group of sounds when written—but then, so is a note in music. And a definition that can apply to more than one thing is a bad definition.

Exclamations like 'ouch!' or 'oh!' are no more than sounds—a sort of primitive music, perhaps. Yet it has been suggested that all language had its beginning in such sounds as these: that the first words were the result of involuntary expressions of pleasure and pain—a contented 'ah!' or a hurt indrawing of breath. Sceptics have dubbed this notion the 'Pooh pooh' theory.

The chief objection to it is that words such as 'ouch' and 'ah' do not fit in with the rest of language—they stand alone; we cannot expect to be understood if we say: 'The Ah went along the ffss!'

Another similar theory (nicknamed the 'Yo-he-ho') is that the grunts and other rhythmic strain-noises which we tend to make while hauling on a rope, or over other heavy work, were the beginning of sea-shanty-style songs and that everyday talk developed from these chants.

Other people have argued that all words are in some way imitative—the 'Bow-wow' or' Ding-dong' theories, which explain themselves. Most words are not apparently imitative, but a large number are: 'quack', 'whinny' and 'twitter' are of the 'Bow-wow' type; 'splash', 'bang' and 'clatter' come into the 'Ding-dong' category.

There are also a number of gesture theories, the best-known of which is the Ta-ta. According to this hypothesis, hand and tongue tend to act together—as when both 'wave goodbye'. To put the point another way: if a man were unable to use his hands, he might, to indicate himself, draw in his lips—and utter accidentally something like 'me' or 'mich' (German) or 'moi' (French). Or, wishing to indicate another person, he might thrust out his lips as pointers and produce 'thou' or 'du' or 'tu' in so doing. (Check your lip movements as you pronounce these words.)

The Bible gives an account of how Adam and Eve named the things about them—but does not indicate why those particular names took the form they did.

If we could believe that all words, even those which do not seem to be names of things—words like 'red' and 'yellow'; and 'but', 'not' and 'with', for instance—developed from names, we should be over-simplifying, but we should be getting somewhere. Bertrand Russell, in *An Inquiry into Meaning and Truth*, accounts for 'not' as the direct experience of inhibition. Similarly, it has been suggested that words like 'but', 'and' or 'with' are the names of certain inward feelings—'but', for example, is supposed to indicate a feeling of resistance. Again, many words which are no longer the names of things that may be seen, smelt, touched, tasted or heard can be traced back to such names—'orange' for example, used as a colour word. Or, take the word 'consider'—not a thing at all—yet it comes from the Latin word for a star, 'sideris'. Can you find out the connection?

There may well be truth in all the theories summarised here; but it seems most unlikely that all words began in the same way, that one of the theories is entirely true, and the rest entirely untrue.

The question 'How did words begin?' has often been asked, but has not been answered here; or, for that matter, elsewhere. However, many verbs are clearly of the 'Ding-dong' or 'Bow-wow' type; and much complex terminology can be traced back to very simple names—but, as has been said, we do not know why the names took the form they did.

A 1 Write down as nearly as you can ten sounds, which are not proper words, but simply expressive exclamations. Begin with those given in the second paragraph of the passage.

A 2 What is the chief argument against the 'Pooh pooh' theory?

A 3 What is a sea shanty? Find an example.

A 4 State in your own words what is meant by the 'Ding-dong' theory.

A 5 The 'Bow-wow' theory says that man imitated the sounds made by various animals. Can you think of an objection to this idea?

A 6 List ten more words which are clearly imitations of sounds—as are those in paragraph five.

A 7 Write down twenty colour words. Mark those like 'orange' which were originally names of fruits or of flowers or of metals.

A 8 Write down ten words which are the names of people.

A 9 Now ten words which are the names of places.

A 10 Finally, ten words which are the names of everyday things—not trade marks, though.

B 1 How does your dictionary define a word? Anyway, what is a definition? Find out the origin of this word. You have already traced back 'consider' to the noun 'star', do the same thing with: 'imagination', 'capacity' and 'language'.

B 2 How many words are there in the English language?

B 3 Name as many countries as you can whose official language is English.

B 4 How many people and how many countries spoke English four hundred years ago?

B 5 Which language is spoken from childhood by the greatest number of people?

C 1 Sea shanties were mentioned earlier. For A 3 you probably gave some such answer as 'A song with a chorus, sung by sailors, while heaving at a rope.' With mechanisation, songs designed to help us work have gone out. Do you see any 'remnants' of this sort of music in the adult world? The real equivalent, though, are the rhymes still chanted by children: those used in picking sides ('dips') and those which accompany skipping and ball-bouncing games. Make a group anthology of one or all of these types of rhyme, explaining the actions which go with the verse.

C 2 Alexander Pope wrote in *An Essay on Criticism*:

'Tis not enough no harshness gives offence,
The sound must seem an echo to the sense:
Soft is the strain when Zephyr gently blows,
And the smooth stream in smoother numbers flows;
But when loud surges lash the sounding shore,
The hoarse, rough verse should like the torrent roar.
When Ajax strives some rock's vast weight to throw,
The line too labours, and the words move slow:
Not so, when swift Camilla scours the plain,
Flies o'er the unbending corn, and skims along the main.

To what extent has the poet been successful in his display of echoing the sense? Treat each of the four relevant couplets separately. Is the notion of the sound–sense relationship an illusion? If we did not know the meaning of the words in the poem, but merely how to pronounce them, would we appreciate the changes of mood, do you think? It is not enough to judge on the basis of someone else's skilful reading—why? In short, is a return to the 'Ding-dong' view of language practical?

C 3 Try to construct a sentence which contains words which can be traced back to any one source. You are most likely to succeed with onomatopoeic words, or with words derived from names.

2

ABC: Hieroglyph or Cattle Brand?

Spoken words came before letters. Pictures may have come before either, for some of the best cave paintings are 25,000 years old. Animals were depicted as pierced by arrows or with wounds 'bleeding' from natural drip points in the roof of the cave—'sympathetic magic' it has been supposed. Still, it would be pointless to assert that these paintings were in any real sense the beginning of writing.

However, one famous cave drawing shows an elephant with his heart 'picked out' (the double meaning is, I think, appropriate). This, the most ancient example of a conventional symbol, that which is still used to represent the heart, is about ten thousand years old. It, too, was part of the killing magic—the wishful confusion of image and actuality.

Other early symbols (found on Mesopotamian pottery) include: *, star, god, or heaven; ≋, water; ◍, earth; ⊠, the deep, heaven. These signs are important in the development of the alphabet, though the idea behind their use appears to have been, again, to make magic, not to communicate with other men.

Anyway, we can normally more easily communicate with others by means of speech: the primary function of writing is to assist our fallible memories. Thus, the earliest examples of writing give details of property. The form used is pictographic.

A pictograph is a drawing of a natural object or man-made article: the symbol recalls the thing itself to mind not its name. The oldest known pictographs, which date from about

3500 B.C., are shown on this page. Here are signs for head, hand and foot, for a threshing sledge and for numerals.

The second stage in the development of writing can be called ideographic. From the many pictographs, stylised symbols (ideographs) come about which are closely associated not only with the object, but with its name.

One side of a limestone tablet found at Kish.

By the third stage, the phonetic, the same ideographs have come to stand for set sounds—syllables, not necessarily words as a whole.

At the fourth stage of development there are letters, derived from pictures, which 'spell out' each word.

Hieroglyphics (sacred carvings) were used by the Egyptians in an inconsistent way: at stages two, three and four (above) at one and the same time. It has been said that their writing was deliberately kept a difficult business, in order to stop anyone outside the powerful body of priests from mastering it. This famous depiction of the name 'Cleopatra' illustrates the point:

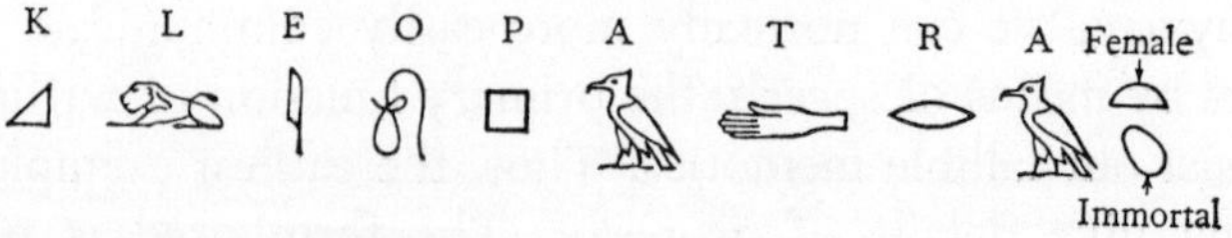

Her name is spelt out—yet, at the end, the loaf of bread tells us (pictorially—there is no attempt to spell out the information) that she was female, and the egg, that she was considered as a goddess. The fact that a symbol might stand for word, syllable or letter (added to the fact that there were some six hundred characters) explains why hieroglyphics seemed undecipherable for so long.

Our own alphabet can be traced from Phoenician times with some certainty. The questionable point is whether it can be derived from Egyptian hieroglyphics. It has been suggested by a number of scholars that cattle brands and mason's marks used by the Semites provide a more plausible beginning. Both possibilities have been allowed for in the chart given on p. 8. It should be understood that the brands and marks had no phonetic implication, that it is their shape only which is significant: it must not be supposed that a 'Mediterranean' mark opposite a certain letter means any degree of correspondence in usage.

The same reservations must be made in the case of hieroglyphics, though here a fairly convincing account of how letters came by their sound can be given. The signs given in the hieroglyphics' column (page 8) were certainly used by the Egyptians. However, when they were used as letters, the Egyptian letter 'A' was not an ox's head; their 'B' not a house, and so on. What may have happened is that the Egyptian ideographs were taken over by tribes who spoke a different language. 'Aleph', in the language of the Semites, meant 'ox'—thus the ox's head became associated with the sound 'A-', and was eventually used for this particular sound whenever and wherever it occurred. It had, then, ceased to be a picture and had become a letter. As it no longer mattered whether it was a recognisable portrait of an animal or not, it was no longer important which way up the sign was written: in fact it tended to be written whichever way seemed quickest and easiest. For some time it lay on its side, but finally turned right over. Our own capital A is probably an ox upside down. But A is the most plausible instance of a hieroglyphic origin.

A 1 In the development of writing, why is the prehistoric heart-shape more significant than the beautifully drawn animals?

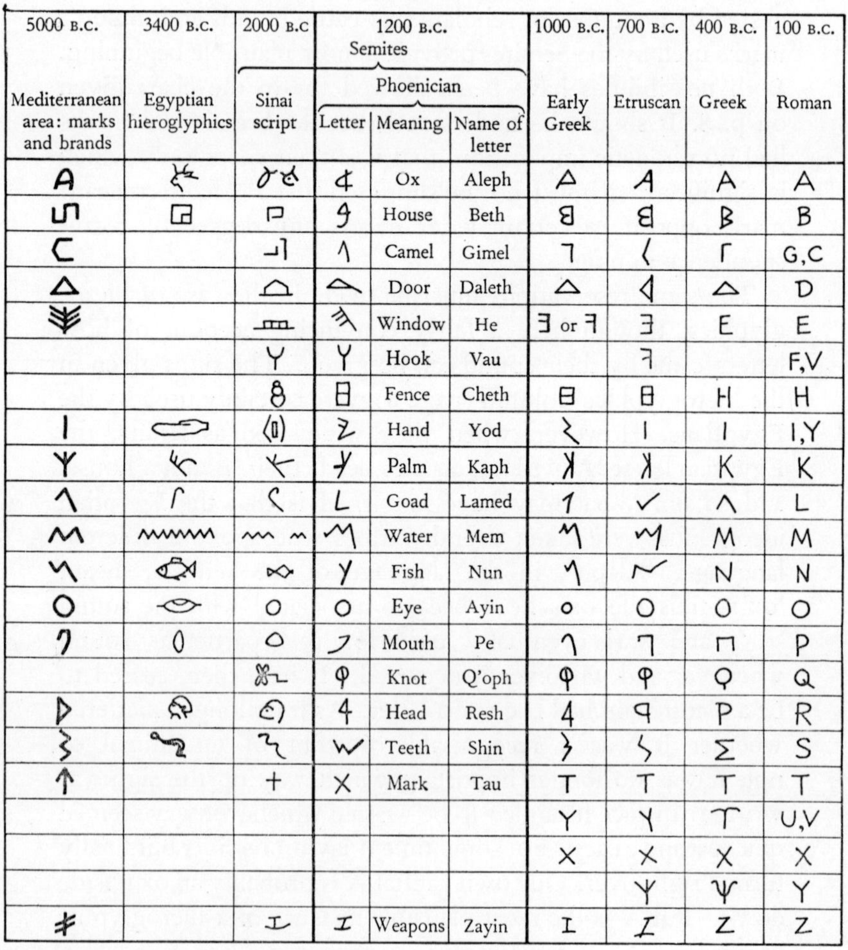

5000 B.C.
3400 B.C.
2000 B.C
1200 B.C.
Semites
Phoenician
1000 B.C.
700 B.C.
400 B.C.
100 B.C.
Mediterranean area: marks and brands
Egyptian hieroglyphics
Sinai script
Letter
Meaning
Name of letter
Early Greek
Etruscan
Greek
Roman
Ox
Aleph
House
Beth
Camel
Gimel
Door
Daleth
Window
He
Hook
Vau
Fence
Cheth
Hand
Yod
Palm
Kaph
Goad
Lamed
Water
Mem
Fish
Nun
Eye
Ayin
Mouth
Pe
Knot
Q'oph
Head
Resh
Teeth
Shin
Mark
Tau
Weapons
Zayin
or
A
B
G,C
D
E
F,V
H
I,Y
K
L
M
N
O
P
Q
R
S
T
U,V
X
Y
Z

A 2 Distinguish between pictograph, ideograph and letter.

A 3 In English, 'A' does not stand for 'ox' nor 'B' for house. Explain this apparent contradiction.

A 4 Which of the sets of characters would take longest to put on paper? Why might the hieroglyphic system nevertheless save time?

A 5 Why were hieroglyphics so difficult to decipher?

A 6 What are the two theories as to how the Phoenicians got their alphabet?

A 7 As far as you can judge from the examples given here, which version seems the more plausible? Why?

A 8 How long has our alphabet existed in approximately its present form? Which of our letters are missing from the Roman alphabet?

A 9 Explain roughly how the use of pictures may have developed into writing by means of letters.

A 10 How might our capital 'A' remind us of the history of the alphabet?

B 1 What is a rebus? When you have found out, try to make a rebus for your own name or for that of a friend.

B 2 Find out the names the Greeks gave to their letters.

B 3 After comparing these names with those the Phoenicians used, say why the ABC is called an alphabet, and what the meaning is of: delta, a delta and a delta-wing—explain the connection.

B 4 Find out as much as you can about the origin of our *numerals*. Construct a chart to show how the numbers 1–9 developed.

B 5 What system of counting was used by (a) the Chinese, (b) the Greeks, (c) the Hebrews, (d) the Romans? Try to write the present year in each of these styles of number-notation.

C Here is a table showing the numerals 1–10 in seven languages:

	English	*German*	*Dutch*	*French*	*Latin*	*Hungarian*	*Welsh*
1	one	eins	een	un	unus	egy	un
2	two	zwei	twee	deux	duo	két	dau
3	three	drei	drie	trois	tres	három	tri
4	four	vier	vier	quatre	quattuor	negy	pedwar
5	five	fünf	vijf	cinq	quinque	öt	pump
6	six	sechs	zes	six	sex	hat	chwech
7	seven	sieben	zeven	sept	septem	hét	saith
8	eight	acht	acht	huit	octo	nyolc	wyth
9	nine	neun	negen	neuf	novem	kilenc	naw
10	ten	zehn	tien	dix	decem	tiz	deg

Note: German v = English f in sound, while z is like English ts.

C 1 Work across the chart carefully, considering each number in turn. Write down the language, which, for this particular number seems nearest to the English version. For example, for '6', you would write 'French'—though, in actual fact, the German 'sechs' is probably pronounced more like 'six'. Also note the one which seems least like English.

C 2 Which language, on this evidence, seems nearest to English?

C 3 Which language seems least like English, and to have little connection with any of the others either?

C 4 Which of these languages, though not as closely connected as Italian, is nearest to Latin?

C 5 You might suppose that the Welsh numbers would be more like those of the English than, say, the German or French ones. Is this so? Attempt an explanation.

3

Man or Parrot?

In 1689 John Locke wrote, in *An Essay Concerning Human Understanding*: 'God, having designed man for a sociable creature, made him not only with an inclination and under a necessity to have fellowship with those of his own kind, but furnished him also with language, which was to be the great instrument and common tie of society. Man, therefore, had by nature his organs so fashioned, as to be fit to frame articulate sounds, which we call words. But this was not enough to produce language; for parrots, and several other birds, will be taught to make articulate sounds distinct enough, which yet by no means are capable of language. Besides articulate sounds, therefore, it was further necessary that he should be able to use these sounds as signs of internal conceptions; and to make them stand as signs of internal conceptions; and to make them stand as marks for the ideas within his own mind, whereby they might be made known to others, and the thoughts of men's minds be conveyed from one to another.'

Nearly two hundred and fifty years later, T. E. Lawrence was making the same point. He was writing of the men of the early R.A.F.: 'Now they were alike-dressed, and all singing "The King of Love My Shepherd Is", with the voices and the pagan enjoyment of their everyday blaspheming. Nor did their minds see any contradiction between their worship and their life. Neither their clean words nor their dirty words had a significance. Words were like our boots, dirty on the field, clean indoors. A daily convention, no index of the fellows' mind. They had not learned to speak' (from *The Mint*).

At first it looks as if John Locke had produced three hundred years ago the definition I needed to begin this book: 'a word is a sound representing an idea in the mind of the speaker which when uttered reproduces that idea in the mind of the hearer'. Yet once the definition is amplified, difficulties become apparent. What, for instance, is meant by 'idea'?—Locke often speaks as if it means mental-image.

It may or may not be the case that, on hearing the sentence 'A pillar-box is red', you have a mental picture of a red pillar-box—but do you have first an image of a pillar-box and then one of 'redness'? And what about 'is'—can there be an image of a colourless pillar-box coalescing with a formless mass of red? And the word 'a'? In fact, the sentence is an examination of a single idea, in the same way that 2+2 is an examination of the idea 4.

Words do convey ideas, though one idea per word is too much to expect—and it is pointless to speculate whether the 'idea' produced in the hearer's mind is the 'same' as that in the speaker's. The only criterion possible is 'does communication usually break down?' If it does not then language is being used successfully. For, in spite of Lawrence, language is largely a matter of convention: one picks up the habit of saying 'of the' on many occasions, finds little temptation to use the construction 'the of'.

There are human beings who cannot speak, but some of them can play chess. The kind of skill necessary to use language is demonstrable without words, in such a matter as playing games: reacting correctly to a given situation within the agreed rules.

A 1 According to Locke, what is the purpose of language?

A 2 What is there to speech beyond the uttering of sounds?

A 3 To what extent does Locke's definition of 'words' hold good?

A 4 Do you agree that it is unprofitable to speculate whether the idea produced in the mind of the hearer is the same as that in the speaker's?

A 5 What other criterion is suggested? What do you think of it?

A 6 What exactly is Lawrence complaining of in the aircraftsmen's attitude to language and life?

A 7 What does Lawrence mean when he says 'they had not learned to speak?'

A 8 Is this a reasonable way of making his point? Why?

A 9 I have suggested that 'a pillar-box is red' is rather like $2+2 = 4$. Do you find this comparison adequate? You will easily find sentences which do more than analyse one idea. Do so.

A 10 What are the similarities and differences between using a language correctly and playing a game according to the rules?

B Lawrence is describing things as they are. Dirty boots (and probably dirty words) are expected on a combat course; clean words and clean boots are the rule in the garrison church. Lawrence is regarding language not as a social phenomenon, but as a means of self-expression. What we call *style* is a mixture of both: the proportions of the 'mixture' will vary from writer to writer.

B 1 To what extent do the Locke and Lawrence passages differ in sound? Which has the more even flow? the more leisurely air? Which comes out in short bursts?

B 2 The tone of Locke's writing is serious and purposeful, the phrasing and vocabulary suited to his subject. What of the Lawrence extract? Are there words or phrases which sound out of place—too formal or too informal for their context? The general tone seems to me to be petulant—do you agree? Relate both passages as they naturally 'read' to the allowances they make for breathing—it is this factor, to a large extent, which determines mood.

B 3 The Authorised Version of the Bible has had a great influence on English style. Can you detect this influence in any of the passages quoted in this chapter (Locke and

Lawrence, above; Wordsworth and Joyce, at the end)? Give full details of your conclusions.

C 1 Suit the sense to the occasion. Frame requests to go into a room: (a) when it is barricaded by a mutinous kid-sister; (b) on the assumption that you are the sovereign requiring technical permission to enter the Parliament building; (c) if you are not sure whether it is an inquiry office or a private apartment; (d) if you are writing to ask if the room (at present locked) could be taken over by the youth club which uses the adjoining hall.

C 2 Suit the style to the hearer. Compare the same news item as reported in *The Times*, *The Daily Telegraph*, the *Daily Express*, the *Daily Mirror*. What kind of reader does the writer appear to have in mind?

C 3 Why—and in what sense—are the following passages 'difficult'?

These beauteous forms,
Through a long absence, have not been to me
As is a landscape to a blind man's eye:
. . . Nor less, I trust,
To them I may have owed another gift,
Of aspect more sublime; that blessed mood,
In which the burthen of the mystery,
In which the heavy and the weary weight
Of all this unintelligible world
Is lightened:— that serene and blessed mood,
In which the affections gently lead us on, —
Until, the breath of this corporeal frame
And even the motion of our human blood
Almost suspended, we are laid asleep
In body, and become a living soul:
While with an eye made quiet by the power
Of harmony, and the deep power of joy,
We see into the life of things.

From *Lines written a few miles above Tintern Abbey*,
by William Wordsworth

'Eins upon a space and a wearywide space it wast ere wohned a Mookse. The onesomeness wast alltolonely, archunsitslike, broady oval, and a Mookse he would a walking go (My hood! cries Antony Romeo) so one grandsumer evening, after a great morning and his good supper of gammon and spittish, having flabelled his eyes, pilleoled his nostrils, vacticanated his ears and palliuned his throats, he put on his impermeable, seized his impugnable, harped on his crown and stepped out of his immobile *De Rure Albo* (socolled becauld it was chalkfull of masterplasters and had bourgeously letout gardens strown with cascadas, pintacostecas, horthoducts and currycombs) and set off from Ludstown *a spasso* to see how badness was badness in the weirdest of all pensible ways.'

From *Finnegans Wake*, by James Joyce

4

That They May Not Understand

John Locke saw language as 'the common tie of society'. For another view, read the story of the Tower of Babel, reproduced from Genesis xi:

And the whole earth was of one language, and of one speech. And it came to pass, as they journeyed from the east, that they found a plain in the land of Shinar; and they dwelt there. And they said one to another, Go to, let us make brick, and burn them throughly. And they had brick for stone, and slime had they for morter. And they said, Go to, let us build us a city and a tower, whose top may reach unto heaven; and let us make us a name, lest we be scattered abroad upon the face of the whole earth. And the Lord came down to see the city and the tower, which the children of men builded. And the Lord said, Behold, the people is one, and they have all one language; and this they begin to do: and now nothing will be restrained from them, which they have imagined to do. Go to, let us go down, and there confound their language, that they may not understand one another's speech. So the Lord scattered them abroad from thence upon the face of all the earth: and they left off to build the city. Therefore is the name of it called Babel; because the Lord did there confound the language of all the earth: and from thence did the Lord scatter them abroad upon the face of all the earth.

A 1 Are the last four words of the first sentence ('and of one speech') merely emphatic—or are two separate assertions being made? Distinguish between 'language' and 'speech' in such a context.

A 2 What was the object in building the tower and the city? Bring out the sense of 'let us make a name'.

A 3 What, by implication, were the Lord's motives in the action he took?

A 4 What is the logical objection to using 'and' to begin each of a series of connected sentences? Can this practice be justified in the present case?

A 5 'Babel' has no known semitic root. From the sense of the passage given here, supply an entry as for a dictionary. Make it as complete as you can: include a definition and an entry under 'origin'.

B 1 In Wyclif's translation of the Bible (1384) the final sentence of the story begins: 'Therefor was callid the name of it Babel, for there was confounded the lippe of all the erthe.' Compare this version with that of 1611 (above) in terms of sentence structure, spelling and what we can reasonably guess about pronunciation. Each account is in its way precise where the other is abstract—give details.

B 2 What is meant by 'lingua franca'? What are the characteristics of pidgin English?

B 3 How many languages are spoken in the world now? You need not include attempts to revive languages which have died out.

B 4 What was the advantage of having Latin as the language of the learned classes?

B 5 Why might it be easier to get everybody to speak Esperanto than, say, English or Japanese? In what ways is Esperanto different from such languages as these? What are the disadvantages of Esperanto? How much progress has it made in what it has set out to do?

C 1 In terms of the cosmology which the Fathers of the Church took over from the Greeks, how high would the tower have to be before it reached heaven?

C 2 It is not clear from the story whether the scattering took place because of the confusion of tongues, or whether the confusion of tongues was a result of the scattering.

Attempt a common-sense account in general terms of factors which promoted movement of population (and thereby divergence of language) in very early times.

C 3

SUM I am a gentleman
ES thou art a fool;
EST he is a crocodile
Sitting on a stool.

This traditional rhyme, included in *The Cherry Tree* by Geoffrey Grigson, seems to be the basis of Bertrand Russell's famous 'irregular verb':

I am firm
You are obstinate
He is a pig-headed fool.

What points does this 'conjugation' make about (a) language, (b) human nature?

Devise 'irregular verbs' of your own. Here are two to begin with:

I am interested in people . . .
I am on a diet . . .

5

Nevertheless, it is English

Der Hund biss den Mann. You can see that the sentence means: 'The dog (hound) bites the man.' Yet it is written in German.

On the other hand, look at this:

Ne wearð wael mare
on þis eiglande aefregieta
folces gefylled beforen þissum
sweordes ecgum þaes þe us secgað bee,
ealde uðwitan, siþþan eastan hider
Engle and Seaxe up becoman,
ofar brad brimu Brytene sohtan,
wlance wigsmiþas Wealas ofercoman,
eorlas ahrwate eard begeatan.

You stand much less chance of making this passage out. Nevertheless, it is English.

A 1 Identify the words with capital letters—they are names of countries or races, and have changed comparatively little.

A 2 Still, the extract really is harder to understand than the German sentence. Not even all the letters will be known by you.

The sign 'þ' is a letter, now disused, for a characteristically English sound—'th'. This is the sound which foreigners have most difficulty in pronouncing. Replace 'þ' by 'th' in the passage—now which words can you pick out?

A 3 Which letter do the *Germans* use where we have 'th'?

A 4 The extract is actually from an account of a battle fought in A.D. 926. See if you can spot three or four words to do with fighting; also the words for 'island', 'before' and 'from the east'.

A 5 Why do people who wish to give an old-fashioned air to their shop or their goods often use 'Ye olde' as part of the title? *Note:* 'the' has never been spelt or pronounced 'ye'.

B 1 How did England get its name?

B 2 Who were the English?

B 3 What happened to the existing inhabitants of Britain, when the English came?

B 4 Which parts of the British Isles had their own language two hundred years ago?

B 5 Which of these languages are still spoken?

The account of how the English drove the British back into Wales is taken from the *Anglo-Saxon Chronicle.*

C 1 Which king began these records? How many years does the chronicle cover? Could the defeat of the British be an eyewitness account? In what obvious ways does the *Anglo-Saxon Chronicle* differ from *Beowulf* the other famous work written in Old English? Try to get hold of and read a modern version such as *Beowulf the Warrior.*

C 2 The most obvious quality of Old English poetry is its emphasis on alliteration. Sir Philip Sidney, in *An Apology for Poetry*, speaks of it as, 'coursing of a letter, as if they were bound to follow the method of a dictionary'; Shakespeare makes fun of it in *A Midsummer Night's Dream*:

> The raging rocks, And shivering shocks
> Shall break the locks Of prison gates:
> And Phibbus' car Shall shine from far,
> And make and mar The foolish Fates.

or

> Whereat, with blade, with bloody blameful blade
> He bravely broach'd his boiling bloody breast.

However, from first to last the English language has been full of it.

Advertisers do not take over poetical devices out of sheer love of art—yet they have made wide use of alliteration, assonance, rhyme and the rest. Why? What is the purpose and effect of these adjuncts of commerce and poetry? Write notes and observations on alliteration in literature and/or in publicity material.

Make a collection of traditional rhyming phrases of the hugger-mugger, helter-skelter, higgledy-piggledy variety.

6

Ties with German and French

You will probably have realised from the work about numbers that the three languages—English, German and French—are connected. Modern English and modern German are most closely allied, as they are both descended from the *same* old language; on the other hand, France is our nearest neighbour and most of our new words have tended to come from that country.

List 1 gives some English words which have been taken over by the French (and, in some cases by many other countries too) together with some French words which have passed into the English language. Notice that words in their new language have often changed their meaning or spelling.

1.

New French words (English meaning given where not obvious)	*New English words* (French meaning is given)
la bière	crayon (pencil)
le bifteck (beef steak)	café (coffee)
le rosbif (roast beef)	vinegar (vin, wine; aigre, sour)
le football	gâteau (cake)
le five o'clock (tea)	encore (again)
le quickstep	porter (to carry)
le pull (pullover)	face (front)
le hockey	bureau (office)
le rugby	umbrella (l'ombrelle, sunshade)
le tennis	robe (dress)

Note: 'new' was a convenient word as a column heading, but some of the words taken from the other language are by no means recent. Which ones do you think have been adopted in the last fifty years?

The second list consists of a few modern German words with their meanings.

2. der Stuhl (the chair); der Vogel (the bird); der Baum (the tree); das Tier (the animal); die Zeit (the time); der Hund (the dog); die Feder (the pen); die Füchsin (female fox); das Loch (the hole); der Zaun (the hedge).

A 1 Explain briefly why there are many similar words in the three languages.

A 2 Give the normal English meaning for each of the words taken from French in list 1.

A 3 Suggest an explanation for the change in meaning of 'l'ombrelle'. (What other use might the English have for a sunshade?)

A 4 On the evidence of list 1 the English have tended to 'export' words connected with—what? Answer the same question with reference to the French.

A 5 Write down the English nouns which, although they have slightly different spellings and meanings, are the same words as the German ones in list 2. *Zeit and Zaun* are the hardest—German 'z' becomes English 't'.

B 1 German influence on the English language remains most marked today in Scotland and the north of England. In some cases a different word from the normal English one is used, in others it is a question of pronunciation.

Find the meanings of: ye ken; lang; gude (German, gut); kirk; mair (German, mehr); burn; gang; bairn (German, gebaren); braid (German, breit).

B 2 Note that 'deer' once meant any kind of animal, 'fowl' any bird, 'meat' included all food, and 'hound' meant any kind of dog. How does the present-day meaning differ from the original Old English and German one?

C 1 The most commonly used English words are Germanic in character. However, is English more like French or German in its grammar—its word order, 'cases' or lack of them, tenses, genders? If you do not know both languages try an unmodified word-for-word translation with the help of a dictionary. Which sounds least un-English?

C 2 Find out the names, times and sorts of meal eaten throughout a day in (a) France, (b) Germany.

7

History and Language:

Britain can hardly be considered as Britain, during the great Ice Ages. Then, most of the country was covered by glaciers, while the habitable part was joined to the continent, with the Thames as a tributary of the Rhine. But, when the climate once again became warmer, the ice cap, which had been nearly 4000 feet thick, melted. Consequently, the lower-lying land was flooded and a completely new coastline came into being—of, more or less, the present familiar shape.

The neolithic men who inhabited the country after this time (some 5000 years ago) are believed to have come originally from the countries south of the Black Sea. There is no record of their language, though their polished stone axe-heads are common archaeological finds.

Although Britain was no longer linked by land to the continent, the British Isles remained temptingly close—history tells of a whole series of immigrations and invasions from Europe.

The Celts came to Britain in the first place in search of territory to replace that which had been taken from them on the mainland. First among them came the Gaels, who gave their language (Gaelic) to Ireland and to the north of Scotland. Then there were the Britons who took over the south and south-east of the islands.

It was the Britons who resisted the Roman landings in 55 B.C. Julius Caesar did not intend to conquer the country, but he believed that men from the British Isles were being mustered on the side of the Gauls against Rome, and wished to spy out the land in case there was need to take more decisive action later. Over a hundred years went by before Britain actually became a Roman colony.

Once there, the legions stayed until A.D. 410 when they were recalled to Rome—for Britain was an outpost of the empire and the Romans were under attack much nearer home.

With the military gone, the British Isles were threatened by warlike Germanic tribes. The Saxon chieftains Hengest and Horsa were invited to land by one of the British kings, but the Saxons soon quarrelled with their allies and gradually possessed themselves of the kingdom of Kent. The date of the landings was A.D. 449—an important one for the English language as the Saxons necessarily brought their own tongue with them: Anglo-Saxon, otherwise called Old English.

There were further landings by the Angles, Saxons and Jutes: the Britons resisted strongly—successfully in Cornwall and Wales.

From about the year A.D. 800 there came more attacks—by the Norsemen who eventually settled in the north and east of England.

The history of England to this point is fixed in our place-names.

A 1 Explain in your own words how the land link between Britain and the continent came to an end.

A 2 How do we decide what is 'history' as distinct from what is 'prehistoric'?

A 3 What first caused the Romans to consider an invasion of Britain? Approximately how long was Britain under their rule and what caused them to leave?

A 4 Draw a sketch-map to show the successive landings from the continent. Give dates where you can. Shade in the areas still held by the Celts after the final landings of the Anglo-Saxons.

A 5 Why are Scandinavian names quite common in the north and east of England?

B 1 The sequence of events which led to the establishment of English as the language of our country is not really so simple as my account has suggested. The date A.D. 449, for instance, is that given by the Venerable Bede, who wrote three hundred years after the event. What linguistic evidence might there be that the Germanic invaders came from three separate tribes?

B 2 'If the Anglo-Saxons had never come we should now be a people speaking a neo-Latin language, very like French'—what assumptions does such a statement depend on? Is there any evidence on this issue?

B 3 In England many Celtic place-names and personal names survived the Saxon conquest—yet, apart from these, probably not more than a dozen Celtic words became part of the English language. How can these facts be accounted for?

C Draw a sketch-map to show how Britain may have looked when it was still joined to the rest of Europe. Use an atlas to help you. Notice how Scotland seems to reach out for Ireland, as does south-east England for France. You can take it that shallow sea areas such as the Dogger Bank were once dry land. What evidence might there be to support the idea that Britain was once attached to Europe in the way shown by your map? Indeed what evidence is there? It is not enough merely to state that books say so.

8

. . . fixed in Our Place-Names

Celtic names remain in the west country especially in Cornwall, Wales and Cumberland; names of hills and rivers are the commonest examples. The rivers Axe, Exe, Usk, Esk all come from the same Celtic word for 'water'. A 'cumb' or 'combe' as in 'Babbacombe' is a deep valley. 'Pen-' (very common in Wales) means a hill; 'Eccles' or 'eglwys' a church.

When the Anglo-Saxons took over much of Britain they referred to Celtic settlements as foreign, using the Old English word 'walh'. Thus 'Wales' means 'the country of the foreigners': the name that the Welsh gave themselves was 'Cymry' as in 'Cumberland'.

There is surprisingly little trace on the map of the long years as a Roman colony, though there are various names ending in -chester, -caster, -cester, all of which come from the Latin word for 'camp' (castra): the half-name 'port' is also Roman. The English called what had been Roman settlements 'streat' or 'strat' after the Roman roads (Latin 'strata', a paved way).

Most of our place-names are Anglo-Saxon. Places ending or beginning with 'burn', 'bourne', 'borne', 'brook' or 'well' were so-called after various names for stream. 'Ford' or 'Forth' were fording places, often named after the river; but not always: Deptford means 'deep ford', Stafford means 'stony ford'.

The most common ending of all is perhaps '-ton' (compare the modern word 'town') as in Aston, Weston, Norton, Sutton, and Milton. 'Ton' is a settlement and Aston is in the east, Weston in the west, Norton and Sutton north and south respectively, and Milton in the middle.

Some other Old English names and their meanings are: -mere = lake; -ham = home; -hope = valley; -wick = farm; -don = hill (compare the words 'down' and 'dune').

The following place-name endings are Scandinavian: -by (as in 'Derby') = village; -fell (as in 'Scawfell') = hill; -scough = wood; -thwaite = clearing; -toft = homestead, or a piece of land.

A 1 Write down three words, taken from the Romans, which appear in our place-names.

A 2 Why are Celtic names found chiefly in the west?

A 3 Say in your own words how (a) Wales, (b) Cumberland got its name.

A 4 Is 'Wales' a Celtic word?

A 5 From which language do most of our place-names come?

B 1 Who were the Celts? Name six areas where people of Celtic origin live today.

B 2 Has your county got a Celtic name?

B 3 Draw a sketch-map to show the Roman roads of Britain. Mark in, using the modern names, as many towns as you can find, which had a Roman camp. Also show five coastal towns which include the word 'port'. Find, too, any places beginning with 'Strat' or 'Streat': are they on Roman roads?

B 4 'Burgh' or 'borough' is the Old English word for 'city': towns like Edin*burgh* must, then, have been important settlements some thousand years ago. Write down the names of six more.

B 5 Write down the name of your town or village. Put by its side the earliest spelling of the name that you can trace. Go on to find out its meaning. (If your own town or village has a meaning which you know without having to do any research, find out the origin of your county town's name.)

C 'Street' is the nearest equivalent to the Roman word for a paved highway. Quite possibly your own home is situated in a 'street', though it is more likely to be in a 'road'. But if you live in a New Town or some other modern estate, you will almost certainly not live in a 'road' or a 'street'—but in 'areas' with names like 'Arkwrights', 'Orchard Croft' and 'Longfield'. These particular examples are taken from the town of Harlow and are based on an old map which marked the fields by name: one belonged to a man called Arkwright, another was a croft (or small-holding) which included an orchard.

This style of naming has revived an old tradition, as in 'the Haymarket' in London, or the 'Bull Ring' in various other cities. Has your own town (or your nearest town) any such road names, which refer to what once went on in what are now ordinary streets?

Street, road, lane, drive, avenue, close, crescent, boulevard, way, walk, grove, chase, terrace, gardens, parade, square, approach, park, drive—what, strictly speaking, is the distinction between these possible names for 'roads'? What is your address? Is part of it included in the above list?

The name of your road may fairly represent what it is now or what it once was: or, on the other hand, it may merely be an attempt to persuade people that it is a good place to live. For example, 'Southview Avenue' would be a fair enough description of a tree-lined road running from north to south; but a 'Paradise Street', or 'Devon Street' next to the gas-works—both these examples are from Birmingham—take some believing in. Which category does your own road come in, factual or would-be-attractive? Make a collection of all the inappropriate street names you come across.

9

Latin Influence Comes Next

England became a Christian country in the seventh century, and Latin words to do with the religion were taken over; examples are sermon, passion, crucifix, saint and miracle.

After the Norman invasion, England had a French ruling class. For three hundred years, French (a tongue derived from Latin) was the official language: spoken by the nobility, by the 'parliament' (which still bears this French name: 'parler' = 'to speak') and in the law-courts.

A On the left are given words of English origin, on the right French ones of the same meaning.

English	*French*
friend	l'ami
child	l'enfant
hunt	chasser
ox	le bœuf
basket	le panier
swine	le porc
calf	le veau
town	la cité
room	la chambre
twelve	douze
go in	entrer
king	le roi
shire	le comté
sheep	le mouton
neck	le col

Write down in each case a word in common use which is taken from the French. For instance, number one would be 'amity', which means *friend*ship. The rest are easier.

B 1 Who was the first archbishop of Canterbury? Why is the head of the Church of England archbishop of this see?

B 2 The Roman Empire was Christian from the time of Constantine and Britain was part of the Roman Empire; were there, then, no Christians in the British Isles before the year A.D. 597?

B 3 'Norman' and 'Norseman' sound alike. What connection was there between the two?

B 4 The Norman kings spoke French. Which other kings and queens has this country had whose native language was not English?

B 5 Which countries have more than one official language? (Postage stamps may help here.)

C 1 Ian, Sean, Evan, Hans, Jean, Juan, Giovanni, Jan and Ivan are different countries' versions of which Christian name?

Name the countries.

Try to find out how to pronounce each of them. What is the position of the tongue when pronouncing a 'j', a 'sh', a 'v', or a 'y' sound?

Allowing for accidental changes in the course of time—the public house the 'Goat and Compasses' was, it is said, once 'God Encompasseth Us' — would you say that all the names from Ian to Ivan were an attempt at uttering the same sound?

C 2 You will have noticed that the words for edible flesh are of different origin from those for the live animals. From this evidence alone what can we tell about the ruling and the ruled classes?

10

A Family of Languages

Most languages of Europe and some of Asia are descended from a tongue which was spoken some 5000 years ago by a neolithic people, living in the Black Sea area. Existing languages and some 'dead' ones have been classified into 'families' according to the degree of resemblance between them. Here is a simplified chart showing the European languages which seem sufficiently like one another to be descended from the same neolithic speech.

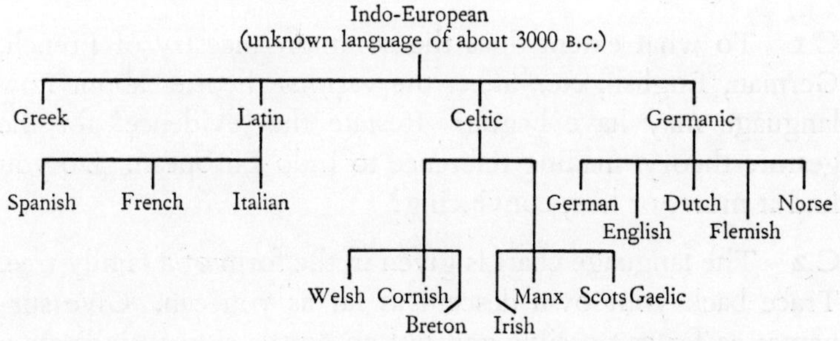

Note: Hungarian, Finnish and Turkish are all connected with one another, but are not directly related to those given above.

A 1 On what grounds have languages been classified into families?

A 2 Which of the languages shown in the chart are (or were formerly) spoken in the British Isles?

A 3 Of the Celtic tongues still spoken in the British Isles which are the most closely related to one another?

A 4 Which European languages have little connection with the ones shown on the chart?

A 5 Use the chart to confirm your findings for the number exercise (chapter 2 C).

B 1 Find out the meaning of 'neolithic' and 'paleolithic'.

B 2 What is a 'dead' language? Give three examples from the chart. It has been suggested that Latin is not so much 'dead' as 'fossilised'—what do you think?

B 3 Britain has languages from the Celtic and Germanic group; which country has languages from the Celtic and Latin groups?

B 4 Roughly half the words in English come from the Latin language (often through French), roughly half from the Germanic. Write a paragraph, explaining why this is so.

B 5 'Most languages of Europe and *some of Asia* ....' Which Asian languages are descended from this same neolithic language?

C 1 To what extent does the common ancestry of French, German, English, etc., affect the various theories about how language may have begun? Restate the 'evidence' for the gesture theory, making reference to Indo-European. Do you find it more, or less, convincing?

C 2 The language chart is given in the form of a family tree. Trace back your own descent as far as you can. Give surnames as far as possible and put an equals sign to indicate a marriage.

11

Old English Words are Short:

The difference in effect between words which come from Old English and those based on Latin can be summed up by this verse:

Scintillate, scintillate, globule orific,
Fain would I fathom thy nature specific;
Loftily poised in ether capacious,
Strongly resembling a gem carbonaceous.

This is far from simple English. In fact three-quarters of it is modified Latin (see A 1 and A 2 for further comments). But words derived from Old English are short, often monosyllables.

Yet short words can be joined together to make longer ones. Hitler, who believed the Germanic peoples to be the 'master race', tried to cut out of the German language all words which came from Latin. When the British called their invention of a new armoured vehicle a 'tank' (a disguise word), the Germans also adopted this term. But, as the word 'tank' comes from the Latin word 'stagnum' 'a pool', Hitler, when he came to power, would have it that a German word should be found. Thus an attempt was made to have it called: *der Schützengrabenvernichtungsapparat* which means 'the machine for destroying the soldiers in the trenches'.

There are many more examples of long words built up from small ones; the Emperor Montezuma's title, for instance: *Montecuzomai-thuica-mina*, meaning 'when-the-chief-is-angry-he-shoots-to-heaven'.

A 1 Put the verse into simple English. Pay attention to the rhythm and you will soon do it.

A 2 All the long words in the verse are based on Latin. What about the few short words, such as 'gem'?

A 3 Why does the German name-finding episode seem ridiculous now?

A 4 Many short and English-sounding names are long Latin ones cut down to size: bus, bike, phone, pram, telly, cab, plane, prom, mob, prole.

Write down the full form of each of the above and try to replace it by words of English origin—the Germans call a bicycle a *Fahrrad*, 'travelling-wheel'.

A 5 But complex words can develop from simple English-elements: cupboard, nickname, pen-knife, breakfast, daisy, nevertheless, goodbye, wireless, doff, railway, evergreen, pinafore, shepherd, atone, business, fortnight, Whitsun, coltsfoot, tadpole, bonfire.

Split up the above into their various parts and add an explanation where necessary.

B 1 What is a monosyllable, a monoplane, a monocle, a monotone? What, then, is the meaning of 'mono'?

B 2 Make up some comic names for chiefs, with a 'translation'.

B 3 Of all the words which most of us have heard, 'antidisestablishmentarianism' is least often used in a normal way. It is often seen, as I have used it here, with inverted commas—how do *they* show it is not used normally? However, it makes its claim as the longest word in the language, so you had better find out what it means. It remains rather a freakish word—what is the effect of 'anti' and 'dis' side by side?

B 4 'Tha knaows', says the Yorkshireman; 'ye ken', says the Scot, who often would not understand him.

'Tu connais', says the Frenchman; 'du kennst', says the German.

What is the distinction between a *language* and a *dialect*? Where does the one end and the other begin? Look both these words up, then decide.

B 5 What is the difference between dialect, slang, jargon, and code?

C Another verse to 'translate':

> There was a young curate of Salisbury,
> Whose conduct was most halisbury-scalisbury;
> He walked about Hampshire
> Without any pampshire,
> Till his bishop compelled him to walisbury.

12

. . . unlike Greek and Latin

Most of our lengthy words—especially those relating to science—are based on Greek and Latin. By means of reasoning, without any reference books even, you can get at the original meaning, which will put you within striking distance of the modern one.

For instance, if the first part of the word 'telephone' (tele-) means 'from a distance', what does the second part of the word mean? Make a list of words containing the syllable 'phone'. What have these words in common?

Or, again, if a 'bicycle' is a thing with two wheels and a 'pedestrian' is someone who goes on foot, what could a 'biped' be?

Here is a list of Greek and Latin words and parts of words—the incomplete ones have a hyphen to show where the rest of the word would be fitted on. They are arranged alphabetically according to meaning.

above	super-	Latin
across	trans-	Latin
after	post-	Latin
again	re-	Latin
against	anti	Greek
against	contra	Latin
before	pre-	Latin (prae)
I believe	credo	Latin
between	inter-	Latin
book	liber	Latin

both	ambi-	Latin
capable	-able, -ible	Latin
to carry	port(are)	Latin
city	urbs	Latin
discourse (science)	logos	Greek
disease	-itis	Greek
distant	tele-	Greek
enough	satis-	Latin
for	pro-	Latin
instead	vice-	Latin
I lead, led	duco, ductus	Latin
life	bios	Greek
light	photos	Greek
I make, made	facio, factus	Latin
not	in-	Latin
out of	ex-, e-	Latin
people	demos	Greek
play, I play	ludus, ludo	Latin
power	kratia	Greek
right	dexter	Latin
ripe	maturus	Latin
ship	navis	Latin
short	brevis	Latin
single	monos	Greek
small	micros	Greek
star	aster	Greek
stone	lithos	Greek
suffering	pathos	Greek
through	per-	Latin
under	sub-	Latin
windpipe	bronchos	Greek
writing, drawing	graphe	Greek
written	scriptum	Latin

Example: You are asked to find a word which means 'to *carry across* country' and are told it has nine letters. You look up 'carry' in the list of Greek and Latin roots and find 'port(are)'. Doing the same for 'across', you find 'trans-'. The hyphen after 'trans' shows that the rest of the word must be tacked on there. Check it for length—answer *transport*. A dictionary, though not essential, will provide an additional check.

A Now, by the same means, find words to fit the definitions which follow:

1. *Above* others in rank. (8 letters.)
2. *Written* on a letter *after* your name has been signed. (10 letters.)
3. To make *new again*. (8 letters.)
4. To oppose someone by *speaking against* him. (10 letters.)
5. You can injure a person's reputation either through malicious talk or by publishing a damaging *book*. Which is libel?
6. *Not capable* of being *believed*. (10 letters.)
7. A *single* standing *stone*. (8 letters.)
8. Picked *before* it is *ripe* (happening before due time). (9 letters.)
9. Performed *between* two halves of a *play*. (9 letters.)
10. Able to use *both* hands as if they were *right* hands. (13 letters.)
11. Having to do with a *city* or town. (5 letters.)
12. A *disease* affecting the *windpipe*. (10 letters.)
13. *Discourse* spoken *before* a play. (8 letters.)
14. If 'roy' is old French for 'king', what is someone who acts *instead* of the king? (7 letters.)
15. This sort of picture has, if you think about it, been *drawn* by *light*. (10 letters.)
16. *Power* of the *people*, a form of government. (9 letters.)
17. *Made* into *stone*. (12 letters.)
18. *Under* the *earth*. (12 letters.)
19. Type of printing begun by *drawing or writing* on a *stone*. (11 letters.)
20. *Study* of *suffering* (causes of diseases). (9 letters.)

B For the next piece of work, the list of Latin and Greek roots will help to some extent, but you will also need an etymological dictionary.

Divide your page into two columns and put each set word in the margin. Head the left-hand column 'origin' and the right one 'explanation'. For example:

	Origin	*Explanation*
aster	Greek word for 'star'	flowers are star-shaped

Now do these: biscuit, curfew, hippopotamus, dandelion, horrid, belfry, biology, navigable, lunatic, Mediterranean, Mesopotamia.

C 1 'The World's Classics', a classical scholar, the number has become a classic, classicism in English verse, the classical revival in English architecture, classical music (in the strict sense), classical music (used loosely), of classic proportions, his third Classic victory.

Explain the use of 'classic' or its derivatives in each of the above phrases. Use whatever means you wish to show that you understand—you might in some cases find it easiest to give an example, in others to refer to an opposite. Is there one general definition which could include all of these instances?

C 2 Your local bank probably has Corinthian columns and rusticated stonework—what? and why? Even the ordinary house of today usually has mouldings (a 'picture rail', for instance) which are relics of cornice or architrave. What exactly is meant by cornice, architrave, capital, entablature, pediment?

C 3 Similarly, as far as most of us are concerned, classical literature is familiar mainly through its echoes. The epic similes of *Paradise Lost*, for instance (see, perhaps, Book 1, lines 192–211). How much of the description clarifies the scene or situation? How much affords us relief—is the 'ornamental' section developed simply for its own sake? Compare with any example from Homer.

In the same way, the Homeric epithet (whereby the dawn is always 'rose-fingered', the sea 'wine-dark', the Achaeans 'long-haired') tends to crop up in unexpected places: 'Come up the drifting sea-dark street now in the dark night seesawing like the sea to the bible-black airless attic over Jack Black the cobbler's shop.' This 'invitation' is from *Under Milk Wood*, written 2500–3000 years after Homer.

Epithets of this sort have compactness to commend them, but in English the effect is always literary, sometimes obscure, and often journalistic.

I really did open a paper 'at random', as they say, and the first sentence which I read was the following one: 'Glad of the chance of a post-Whitsun beano, delegates to the N.U.T.'s Special Conference on May 23 cheerfully filed into Westminster's Central Hall, which they considered a spartan place after Blackpool's Winter Gardens.'

If we consider 'beano', 'delegates', 'Conference', 'filed', 'Hall', 'Gardens' as the key words, how much of the descriptive matter comes before the words? What proportion after? Rewrite this sentence in a more leisurely narrative style which corresponds more nearly to the spoken language.

13

Socratic Method

In Plato's dialogue known as *The Meno*, Socrates argues that all our learning is, in fact, recollection. The soul is immortal, has been often born and knows the Form or principle behind everyday reality. Socrates seeks to prove his point by eliciting in his usual question-and-answer way a geometrical theorem from a boy who knows no geometry. Here is an extract:

Socrates: Are not these four lines drawn of equal length, these, which enclose the square space, *BDWU*?

Boy: They be so.

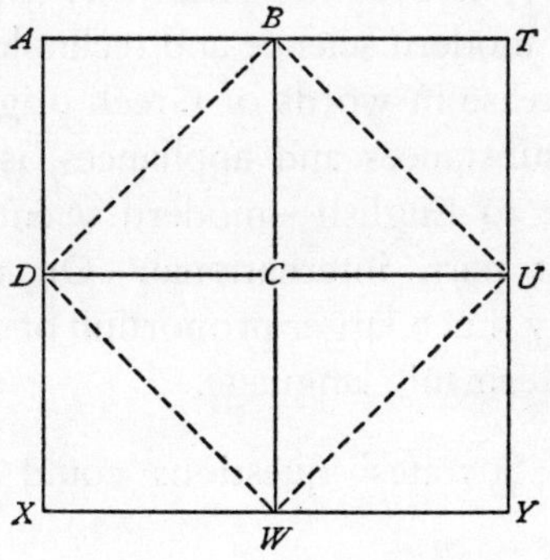

Socrates: Now consider how large this square is which is enclosed by those four lines.

Boy: Why, I do not know.

Socrates: Are not those four squares, *ABCD*, *BTUC*, *CDXW*, *WYUC*, cut each of them in half by these four lines, *BD*, *BU*, *DW*, *WU*, drawn within them; or are they not?

Boy: They be.

Socrates: In the square, *ATYX*, how many spaces are there then, as large as the space *ABCD*?

Boy: Four.

Socrates: And how many such in the square *BDWU* from which half of the other is cut off?

Boy: Two....

(Which leads to the conclusion 'from the diagonal of a square may be drawn a square twice as large as is the square of which it is the diagonal'.)

As can be seen from this closely reasoned dialogue (written in the fourth century B.C.) the Greek mind and, therefore, the Greek language were well suited to precise statement and exact distinction.

When one considers also that the language has an infinite capacity for forming compound words and for producing neat derivatives (where, for instance, the appropriate suffix can determine whether the word is subject or object or process or possibility) it becomes clear why Greek became the root language of modern science and technology.

The rapid increase in words of Greek origin, to match the new processes, substances and appliances, is not surprising; nor is it peculiar to English—modern scientific terminology is, for the most part, international. On the other hand, English has a very much larger proportion of *Latin* derivatives than any other Germanic language.

A 1 Which of Socrates' questions could justly be called 'leading'?

A 2 Which of the boy's answers show insight into the problem?

A 3 Only part of the proof is given here. Which necessary stages of the demonstration would seem difficult for the boy to arrive at without external prompting?

A 4 A mathematical point, once grasped, does give one the sensation of recognising something previously known. Can you explain this 'feeling'—or are you driven to accept Socrates' views on rebirth?

A 5 Which kinds of knowledge have nothing of this 'remembered' feel about them and so would not lend themselves to a demonstration of this kind?

B 1 Etymology has its interest and does help to 'anchor' meaning—but how convincing do you find an appeal to etymology on the following lines?

(a) 'Education' means to draw forth (knowledge), so all teaching must be done by the Socratic method.

(b) 'A post-graduate student' can only mean a dead graduate.

(c) *X:* I've had a hectic time lately.
Y: Wait here. I'll get the ambulance.

(d) 'Telegenic' if it means anything means 'born from a distance'.

B 2 Why *is* the Latin element in English so very much larger than that in German and Dutch? Some notes towards a possible answer are given below. Fill in the details.

Norman conquest...French a ruling language...translations from French literature...Latin words with endings dropped taken over from the French...pattern whereby words could be taken directly from Latin.

C To demonstrate the versatility and precision of Greek, look up (preferably in the *Shorter Oxford Dictionary* where they are successive entries) the following words:

Anthropic	Anthropolite
Anthropocentric	Anthropological
Anthropogeny	Anthropologist
Anthropoglot	Anthropology
Anthropography	Anthropomancy
Anthropoid	Anthropometry
Anthropolatry	Anthropomorphic

Anthropomorphism
Anthropomorphite
Anthropomorphitism
Anthropomorphize
Anthropomorphology
Anthropomorphosis
Anthropomorphous
Anthroponomical
Anthropopathy
Anthropophagi
Anthropophagous
Anthropophagy
Anthropophobia
Anthropophuism
Anthroposophist
Anthroposophy
Anthropotomy
Anthropurgic.

How many of the words are of nineteenth-century origin? How many of these words have you ever heard spoken? How many do you recall having seen in print outside the pages of a dictionary? How true is it to say that the few that are in general use are those that have been in the language longest?

14

Look It Up

The Shorter Oxford Dictionary ('shorter' than what?) has upwards of 2500 pages, each with three columns of print; each column has about eighty lines, at eight words to the line. There are about forty-five main entries per page. To give an idea of the book at work, the dictionary's handling of the word 'pink' is reproduced below in a slightly simplified form —information which depends on knowing certain abbreviations has been omitted.

Pink 1471. (apparently adoption of Dutch *pincke*, small sea-going ship; ultimate origin unknown.) A sailing-vessel; originally one of small size, flat-bottomed and having bulging sides; later applied to warships, etc.

Pink, *penk* 1490. (Origin obscure.) 1. A minnow. Now *dialect*. 2. A young salmon before it becomes a smolt; a samlet, parr 1533.

Pink 1512. 1. (Said to be a nasalized modification of *picken*, PICK.) A hole or eyelet punched in a garment for decorative purposes. 2. A stab with a poniard, etc. A shot-wound 1885.

Pink 1573. (Origin obscure.) A 1. General name of various species of *Dianthus*, a garden plant with very numerous varieties, having pure white, pink, crimson, or variegated sweet-smelling flowers. 2. *figuratively* The 'flower' of excellence; the embodied perfection (*of* some good quality) 1592. The most perfect condition or degree *of* something; the height, extreme 1767....

[There follow examples of the word 'pink' used in combination with another, to connote particular flowers; also illustrative quotations, as: 'Nay, I am the very pinck of curtesie' Shakespeare; and one phrase: *In the pink*, in perfect health (*colloquial*).]

...B 1. Of the colour of the pink in its single natural state; of a pale or light red colour, slightly inclining towards purple; of a pale rose-colour 1720. 2. Applied to the colour of a hunting-coat 1857...[Examples follow, as above].

Pink 1634. (Origin obscure.) A yellowish or greenish-yellow pigment or 'lake' obtained by the combination of a vegetable colouring matter with some white base, as a metallic oxide; as *Italian pink*.

Pink 1540. (=Dutch *pinken* to shut the eyes. History unknown.) Of the eyes: To be half shut. Of a person: To blink or wink; to look slyly.

Pink 1920. (Echoic.) Of a motor-engine: To 'knock'.

A 1 The system of alphabetical entry applies, of course, to the dictionary as a whole; but when the same spelling may have different meanings, as above, what seems to be the principle behind the arrangement? What about the last entry but one? (N.B. words which can be noun or verb have been included by me only as nouns.)

A 2 How true is it to say that the earliest uses of the word 'pink' are obsolete? Be careful about the third entry.

A 3 Go through the list, classifying each entry as follows: obsolete, existing only in combination with another word, present-day (try to assess whether the use of the word is increasing or diminishing).

A 4 What is disappointing about the entry for 'pink'? i.e. what information do even the scholars lack?

A 5 What evidence is there that the pronunciation of 'pink' has undergone changes from time to time?

B 1 Make a similar study of the word 'clutch'.

B 2 How many of the functions set out below can a dictionary perform? Say to what extent a dictionary will be useful in each case and in what circumstances.

To show one how to spell; supply the exact word for a given meaning; increase one's vocabulary; convey a word's pronunciation; give *one* alternative word for the same idea; show the origin of the word, the opposite of the word; lead one to words of similar origin; teach the history of the word; convey its grammatical structure, its position in a sentence, its meaning.

B 3 Are there other reference books which can go over the ground not thoroughly covered by a dictionary? See your answer to B 2. What, for instance, is a thesaurus, a lexicon?

C 1 Whether you know German or not, look up 'der Zug' in the largest German dictionary available. How many meanings are noted? What do they have in common?

Such a range of meaning is making one word go too far for English tastes. Many of our English words which appear to have different meanings are actually different words which have come to be spelt the same—laziness of a different sort. Or, say, a noun has become used as a verb. Thus we have three categories: words which have stretched their meaning, different words which sound and look the same, words which have changed their grammatical function. Find an English word such as 'pink' which has a number of entries or variations—into which category does each different meaning come?

C 2 Some words have a very limited range, are confined perhaps to a particular occasion or to a particular social group. Nevertheless, the words themselves may be widely known: the word 'crucifix', for example, which can be used only with a religious connotation. On the other hand, 'cross', its synonym, ranges from religion to roads. Again, 'corbel' is a word restricted to the discussion of architecture, but 'boss' (which can occur in a similar context) has another, more familiar, application.

Consider the following words: disc, garter, advocate, gear, liege, endorsement, abstract, ablutions, hip, order, interest.

Give instances of their present-day use and write notes on each word, indicating its range. Does the word, or one of its senses, form part of a specialist vocabulary connected, for example, with sport or with one of the professions?

15

Names: of that Person, Place or Thing, Only?

His Eye might there command wherever stood
City of old or modern Fame, the Seat
Of mightiest Empire, from the destind Walls
Of *Cambalu*, seat of *Cathaian Can*
And *Samarchand* by *Oxus*, *Temirs* Throne
To *Paquin* of *Sinæan* Kings, and thence
To *Agra* and *Lahor* of great *Mogul*
Down to the golden *Chersonese*, or where
The *Persian* in *Ecbatan* sate, or since
In *Hispahan*, or where the *Russian Ksar*
In *Mosco*, or the Sultan in *Bizance*,
Turchestan-born....

Taken from the vision of Earth's kingdoms and their glory, revealed by the archangel Michael to Adam, in Book XI of *Paradise Lost*

Name is pounded on name here to give an impression of hugeness and exotic grandeur. How does the use of capital letters in these verses differ from that which you would expect to find in a present-day novel? What are the modern uses of italics? In the passage one noun only does not have a capital letter—why I do not know—see if you can pick it out. How many names of towns or countries are there here? How many of the words appear to be titles of rank? 'King' is the most obvious. What examples of unfamiliar spelling can you find here?—these are not necessarily because the work was written three hundred years ago: Milton had individual views on spelling (see Section C).

The Germans still use capital letters to pick out all nouns (names of things): *der Stuhl ist neben der Tür*, 'the chair (stool) is near the door'.

In modern English only 'proper' nouns are given capital letters. In theory a proper noun is the name of one person, place or thing—and of that person, place or thing only.

This remains largely true in the case of place-names, in spite of the fact that some places do bear what appears to be the same name as others. Names that are now the same may once have been different. There are a good many places called 'Aston', for instance, dotted about England—but 'Aston' may mean 'east village' or it could be 'the farm where the ash tree grows', or the first part of the word may be someone's name. You can find out *which* by checking the old spelling of the names.

Place-names in America, though, are rather different: many are named after people, some are names from the Old World with 'new' added (for example, New York or New Orleans), some merely a straight borrowing from England or other countries—Boston or Birmingham, for instance. 'Cheating', this last case, and in a way contradictory too, for the Birmingham in the U.S.A. never was and never could be the homeland of Beornmund's people, which is the probable meaning of the word. However, the 'one name–one thing' rule holds good if we add the state or county to the smaller of the towns in question, making a sort of double name, as: Birmingham, Alabama; Boston, Lincs.

Christian names have not stuck to the above rule—as any Tom, Dick or Harry knows; but it is worth inquiring to what extent they ever did.

Old English names consisted of two parts, as in: Alf-red, Athel-stan, Ed-win, Ethel-bert, God-wine. Each half of the name had a definite meaning. 'Godwine', for instance, is 'lover of God'. The various possible halves of names were in course of time changed around, the meaning being carefully considered. So that the family history could be easily recalled, descendants generally had the same initial letter as their forefathers.

After the Normans had conquered England, the Old English names dropped out of use, except for those which happened to be names of saints. Biblical names became popular from this time on, especially 'John' and 'Mary'.

In the eighteenth century names such as 'Mary' and 'Helen' tended to become 'Maria' or 'Helena', after the Latin pattern. They seemed smarter to people that way. Also at this time certain Old English names were reintroduced, such as Edgar, Edwin and Alfred.

Books have revived old names and even produced new ones; in the nineteenth century, Wilfred, Roland, Roger and Ralph were restored to popularity in this way.

A 1 How true is it to say that a place-name is the name of that particular place and of no other?

A 2 Small children are sometimes indignant when they find out that their names, John and Mary, say, are not their exclusive property. Attempt to explain, as to a child, why this is.

A 3 What evidence is there that the original meaning of place-names, such as Birmingham or Boston, was forgotten by the seventeenth and eighteenth centuries?

A 4 At what period was special importance given to the meaning of a personal name?

A 5 When and why did the taking over of other people's (unaltered) names begin?

B 1 What motive do you think immigrants would have in naming a new settlement after an old town?

B 2 Which people are these places named after: America, Virginia, Georgia, Maryland, Alberta, Pennsylvania?

B 3 The name 'Pamela' has an interesting history; find out what you can about it.

B 4 In 1938 the names Gary and Shirley were very popular at christenings in the large towns of Britain. Why?

B 5 Carry out a small-scale survey. Which boys are named after their father; how many have their father's first name as their own first name, as their own second Christian name? Do the same in relation to girls and their mothers' names. What are your conclusions? There are other, more public, influences at work. You may be able to find reasons why a number of children of the same age have the same slightly unusual Christian name.

C 1 If Milton had written *Paradise Lost* in his youth he would probably have used a foreign tongue. At the age of nineteen amid Latin verses he apologises to English for his rare use of it:

> Hail native language that with sinews weak
> Didst move my first endeavouring tongue to speak,
> And mad'st imperfect words with childish tripps,
> Half unpronounc't slide through my infant-lipps,
> Driving dum silence from the portal dore,
> Where he had mutely sate two years before....
> I pray thee then deny me not thy aide
> For this same small neglect that I have made.

Milton's spelling in manuscript has an abundance of terminal e's: *sate*, *aide*, *howre*, *heavie*, *devoure*, *vaine*, *drosse*, *losse*, *gaine*, *greedie*, *blisse*, *kisse*, etc. But in the case of *Paradise Lost* his unorthodox spelling seems to have had a system and a purpose: to vary the emphasis through changing the sound. Thus, he would set down the pronoun 'me' as '*me*' or '*mee*'—would lengthen the vowel to stress the word.

Unorthodox views in a different direction were put forward by James Howell, a contemporary of Milton:

'Amongst other reasons which make the English language of so small extent, and put strangers out of conceit to learn it, one is, That we do not pronounce as we write, which proceeds from divers superfluous letters, that occur in many of our words, which adds to the difficulty of the Language: Therfore the Author hath taken pains to retrench such redundant, unnecessary Letters in this Work (though the Printer hath not bin so carefull as he should have bin) as amongst multitudes of other words may appear in these few, *done*, *some*, *come*: Which though wee, to whom the speech is connatural, pronounce as monosyllables, yet when strangers com to read them, they are apt to make them disillables, as do-ne, so-me, co-me; therefore such an *e* is superfluous.

'Moreover, those words that have the Latin for their originall, the Author prefers that Orthography rather than the French, whereby divers letters are spar'd, as *Physic*, *Logic*, *Afric*, not *Physique*, *Logique*, *Afrique*; *favor*, *honor*, *labor*, not *favour*, *honour*, *labour*, and very many more, as also he omits the Dutch K in most words: here you shall read *peeple*, not *pe-ople*, *tresure*, not *treasure*, *toung*, not *tongue*, etc. *Parlement*, not *Parliament*, *busines*, *witnes*, *sicknes*, not *businesse*, *witnesse*, *sicknesse*; *star*, *war*, *far*, not *starre*, *warre*, *farre*, and multitudes of such words, wherein the two last Letters may well be spar'd: Here you shall also read pity, piety, witty, not piti-e, pieti-e, witti-e, as strangers at first sight pronounce them, and abundance of such like words.' (From *English Examined*, a collection of seventeenth- and eighteenth-century views on the English language, by Susie I. Tucker.)

For the origin of the redundant 'e', see chapter 24.

How far has modern spelling followed the pattern laid down by (a) Howell, (b) Milton?

Would you yourself go so far as to say that, for example, 'Wednesday' should be spelt 'Wensdy'; 'been', 'bin'; 'breakfast', 'brekfus'? In other words, are your sympathies with the phonetician (who would like an accurate representation of the sound) or with the etymologist? That is to say, do you think it at all important that we should be able to trace earlier meanings from the look of the word—*Woden's day*, or *break fast*? Take any page from a modern novel. Using present-day pronunciation as a guide, 'simplify' the spelling. Do any grammatical ambiguities arise (for example, the -s ending for plural noun and abstract noun)?

C 2 The poet, W. H. Auden has said that the art of giving a name is akin to that of writing a poem; that there is an indefinable appropriateness in an apt name—for we soon notice if someone's name does not 'fit' ('Oh, he's not a Brian', etc.). What do you think? Which of you are satisfied/dissatisfied with your own names? What would you change them to? Do you associate certain qualities with certain names—intuitively or on rational grounds?

16

Surnames—Much More Informative than Christian Names

While the reasons for choosing certain Christian names are interesting enough, there is not usually nowadays a family tradition about such things, nor is there much regard to the actual meaning of the names.

Surnames are much more informative.

Ask anyone to pick the three most familiar surnames: 'Brown, Smith and Jones', will very likely be the answer.

Yet these three names came about in quite different ways. Which of them comes from a trade or job? Which is more like a nickname? Which is named after a parent? A more obvious example is 'Johnson' (son of John).

TRADE NAMES

Besides the obvious ones: Butcher, Baker, Carter, Cook—see how common nouns have earned a capital letter by being applied to one person—some less simple cases are: Faulkner (forester), Bannister (cross-bow man), Barker (tanner), Fletcher (maker of bows and arrows), Latimer (latiner—someone who could translate English into Latin), Spencer (steward or dispenser), Spooner (roof-maker), Warner (warrener—a gamekeeper).

NICKNAMES

Black, Brown, Long, Lang (tall), Strong, Strang (notice the different forms of the last two—a question of pronunciation), Little, Broad—all these mean what they say. Some others are not so obvious: 'Lock' would refer to a prominent piece of hair, 'Dent' to a tooth (through the French word); while it is surprising that Cameron, Campbell and Kennedy have survived as names, for their meaning is very unflattering. Look them up!

'SON OF —'

Johnson and Williamson are unmistakable; Dawson, Hodgkinson, Henderson, Hobson, Dickinson look puzzling at first. Which Christian names do they come from? ('Hob' was a pet name for 'Rob', 'Hodge' is similar).

'Mac-' means 'son of'; so does the Welsh 'Ap-'. The names Rice, Price and Rees come from the earlier form 'Ap Rhys' which means 'son of the ruler'.

A Go round the class trying to see which of the surnames are occupational in character, which mean 'son of' (or 'belonging to'), which are nicknames—and which type of surname I have not dealt with. Why?

B As familiar as our own names are the days of the week and the months of the year. The days of the week are Old English. Here are their early forms: Monandaeg (moonday), Tiwesdaeg (day of the war-god), Wodnesdaeg (of the god Woden), Thorsdaeg (Thor's day), Frigedaeg (day of the goddess Frigg), Saeterndaeg (Saturn's day), Sunnandaeg (sunday).

You can see now why 'Wednesday' is still not spelt as it is pronounced, but who were Woden and Thor?

The days of the week are Old English, but which language do our months come from?

January is named after Janus—why? June is named after Juno—why?

Find out why the months named after the Latin numbers 7, 8, 9, 10 are, in fact, respectively the ninth, tenth, eleventh and twelfth in our calendar.

The days of the week are all named after gods; the months are a mixture—which of them are named after people, which after gods, which after festivals, which after seasons, which from numbers?

C Names of places; people, and some things that are thought of as being in a way like a person (e.g. countries, rivers, ships); months and days—all these are proper nouns and are written with a capital letter.

But some proper nouns have, in the course of time, lost their capital letters—oddly enough a sign of fame; for some common everyday things have taken their names from people and places. Using a good dictionary, find out the origin of these common nouns—many of them may also be used as verbs, though: boycott, canter, cardigan, currants, guillotine, guinea, quisling, sandwich, tarmac, wellingtons.

Using an encyclopaedia, find out more about three of the *people* who gave their names to these things.

17

Word for Word

In the back of Mark Twain's book *A Tramp Abroad* may be found an amusing, but unfair essay on the German language. It also has pictures of such things as a long German word coming at you over a hill and stretching back for miles in perspective.

One of the tricks the writer employs is to translate literally (word for word) from German to English. The chief figure of the 'translated' story is 'das Fischweib', 'the fish-wife', Mark Twain translates it. He made that choice of subject, because 'das Weib' is a neuter word and so he could translate the pronoun form as 'it', thus adding to the fun.

In the German language the structure of the word, not the nature of the thing, determines its gender or sex. So the word for a pen (die Feder) is feminine, though the thing itself has no life; and a 'nail' (der Nagel) is masculine, while a girl (das Mädchen) is neuter!

With the aim of illustrating a word-order different from our own, I have done the same sort of German–English 'translation'; in this case from the preface to a German dictionary. The passage deals with English pronunciation (as compared to German).

The Pronunciation of the English distinguishes itself from the german chiefly thereby, that the Englishman with Speech the Tongue sinks and backdraws, so that she flatter in the Mouth lies. The Underjaw becomes a little to front pushed and the Lips become much less used, above all not forwards tilted, at most middling rounded.

It is only fair to say that no language will translate into any other in this word for word way.

A 1 How does Mark Twain make fun of the German language?

A 2 Why are his methods unfair?

A 3 In the 'translation' the German system of capital letters has been retained. What is the purpose of the 'extra' capitals? (This was mentioned earlier in the book.)

A 4 From the clause 'so that she flatter in the Mouth lies' (half-way through the 'translation') find an example of (a) a verb placed in an 'un-English' position, (b) 'illogical' gender.

A 5 Rewrite the hints on English pronunciation, putting them into natural-sounding English: you will need to alter sentence-order and expression a good deal.

B English mothers sometimes refer to their babies, in an affectionate way, as 'little ducks'; in the same way, French mothers may call infants 'petits choux'. Yet 'un petit canard' and 'little cabbage'—the two expressions taken over word for word from one language to the other—would sound equally silly, applied to children. The nearest translation of 'petit chou' would probably be 'duck', though 'chou' means a cabbage. The literal *meanings* of the two expressions are different, but they are *used* in the same kind of way. Expressions like these, which cannot be translated exactly, are called idioms.

Thus, the Germans talk of 'drawing your hide over your ears'—which means much the same as, and makes rather more sense than, our own 'paying through the nose'. The French use a different verb to the English in their references to the weather (*il fait beau*, it makes fine); we complain of 'being under the weather' when we are not, of course, under it and when we are not really talking about the weather anyway.

Now give the meaning of each part of the strange conversation set out below. Say why a word-for-word translation is not likely to be a good one—point out how what is meant differs from what is, on the face of it, said. If you are studying a second language, you might even try looking up each word in a French or German or Latin dictionary, so as to make your own very bad literal translation. You yourself will probably be able to see that the resulting piece of work sounds awkward and unlikely—but show it to your language teacher and see what he or she has to say! If you are an advanced student, try to find the equivalent idiom.

'In the small hours, you say?'

'Go on!'

'Well, he wasn't pulling my leg, I'm sure.'

'They've given him the sack, then?'

'Yes, I'm afraid they have and I must say I'm pretty down in the dumps about it.'

'All of a sudden, wasn't it?'

'And his future at stake at that.'

'Is he throwing in the towel?'

'No, he reckons it's red tape as usual.'

'Either that or someone got out of bed the wrong side.'

C We often do not mean what we say. If we 'stop dead' we don't actually die, nor do we nowadays get our sack returned to us when we lose a job (we do not really *lose* it either); and a teacher who says 'mind you don't strain yourself' is almost certainly suggesting that you work harder.

Sarcasm is a statement which is made in such a way that it is clear the exact opposite is meant. Make sarcastic reference to: someone who is always miserable; very dirty hands; news that comes too late; bad weather; the work you like least.

The feeblest form of sarcasm generally has difficulty in showing that it is not meant to be taken seriously. If, in tackling any of the above, you have ended by adding 'I don't think', that, I'm afraid *is* pretty feeble.

Sarcasm is spoken, *irony* is its written equivalent. The two words mean much the same—but irony does not depend upon a tone of voice, its meaning is made clear from the words themselves. Compare the second stanza of *Sing A Song of Sixpence*.

A particular form which is easy to spot, though, is so-called 'dramatic irony': the audience is in the secret while the characters on the stage are not; some statement is made which they believe in, but which we know to be the opposite of the truth. In *Macbeth* for instance, Macduff is told that he has suffered nothing at the hands of Macbeth. He agrees, but we know that Macduff's whole family has been wiped out at Macbeth's command.

Be on the look-out in plays and novels for examples of dramatic irony. Can you think of any off-hand?

18

A Basic Sentence Added To

Here is a normal-sounding English sentence:

The centre forward booted the ball into the top of the net.

This sentence could be considered to answer the following questions: Who did it? What did he do it to? Where did it go? Write down in the briefest possible way the answer to each of these questions.

You should find that you have copied the original sentence which is of the type most usual in English.

Information, then, comes to us in the following order: Who? Did what? To what? How, Why, When or Where?

It saves time to give each section of the sentence a label: it is obviously helpful for a teacher to be able to tell a pupil that there are, say, no verbs in what he has written.

The normal English sentence-order, then, is:

subject	*verb*	*object*	*adverb*
(Who?)	(Did what?)	(To what?)	(How, why, when or where?)

A pupil has written the 'sentences' which are given below and you have to mark them. Write by the side of each one the 'label' for the piece missed out.

The centre forward the ball into the top of the net.
The centre forward booted into the top of the net.
The centre forward booted the ball.

Which of the 'sentences' makes complete sense as it stands? Which, then, is a sentence?

So the adverb is in a way inessential; yet it is the phrase 'into the top of the net' which gives the sentence what interest it has.

The same is true of *adjectives*; for our basic sentence can, naturally, be added to. The centre forward could be burly and bustling or weakly and knock-kneed—all of which descriptive terms answer the question 'what kind of?' and are adjectives.

A 1 What is the most usual type of English sentence?

A 2 If you wanted to pick out the subject of a particular sentence, what question would you ask yourself?

A 3 And what if you wanted to find the 'object'?

A 4 What advantage is there in knowing the meaning of words such as 'verb' or 'subject'?

A 5 Which part of the pattern sentence seems inessential—why is it, nevertheless, important?

B 1 (a) Go through the alphabet mentally, finding for each letter an adjective that is appropriate to the centre forward (for example, artful, bustling, crafty, etc.). How far do you get before you are stuck? Write down your considered selection.

(b) Give him a string of epithets, being guided by sound as much as by sense; for example: 'The baldy-pated, goaly-beating, candle-eating, muddy, snub-nosed, rose-scented centre forward....' Load the whole sentence with extravagant descriptive words.

(c) Invent an alphabetical paragraph or two—'A big cat doesn't even feel', etc. Try doing it quickly. You will probably find that you have adhered fairly closely to the basic sentence-order, for that is the sequence of ideas that the English language handles most easily.

B 2 You have been asked to take part in a survey on 'Schools and Pupils'. Use complete sentences, but underline the words which give the vital information:

When did you first attend school?

How do you travel to your present school?

What would you say was the purpose of compulsory education?

Where was the first school you ever attended?

Whether you have taken these questions seriously or not, all your answers will contain adverbs in one form or another—they may be one-word adverbs, adverbial phrases or adverbial clauses. Check which they are.

Adverbs, then, answer the questions: When? How? Why? Where?

For your answer to the second question you may have put: by bus, on foot, slowly, quickly—or many others. Adverbs most *commonly* answer the question *How?*

'Tom walked—how?' Complete the sentence; 'hurriedly' would fit. How many more suitable adverbs can you find in five minutes?

C 1 'The Parson's Cat'—a well-known parlour-game can easily be adapted for school use. A letter is selected and the statement 'the parson's cat is a — cat' is written on the board. The statement must be completed by the insertion of an adjective (or adjectival phrase) which begins with the chosen letter. If it were 'A' some answers might be: artful, angry, artificial, Australian, ant-eating, anxious. The answers must make sense, although they may sound as unlikely as you please. Those who fail to answer within the agreed time are to stand up, until there are more players 'out' of the game than 'in'—at which point those out of the game sit down while the survivors stand.

A more difficult version of the game calls for prepared cards; one pack for the vowels; one for the consonants, counting 'y' as a consonant and omitting 'x' and 'z' as too difficult. First time round the consonant pack is cut for the initial letter. Second time round the vowel pack is cut to provide the second letter (unless the consonant is 'Q', in which case the vowel pack provides the third letter).

C 2 Look at these two groups of words:

1. lark, lick, like.
2. car, cook, cow.

Within each group, the consonants (the sounds 'l-' and 'k-') are pretty much the same, but the vowels seem different. Yet, the vowel-sound of the words 'like' and 'cow' is made from a combination of the two previous words. The first two words in each group have a single vowel-sound in the middle, while the last of each set of three has a diphthong—a double vowel-sound.

In the first case the 'ie' sound of 'like' is made up of the 'ar' sound of 'lark' and the 'i' sound of 'lick'. LIKE-lar-ick. Do not sound the 'r', though.

Work out the second set in the same way.

With help if necessary, pick out which of the following words have simple vowels and which diphthongs: lee, lid, led, lad, lard, law, ludo, laid, load, lie, allow, loyal, learn, lair, lore, lure.

It is said that French vowel-sounds are 'pure'. What does this mean? Read Wilfred Owen's poem *Strange Meeting*, which has a strange echoey sound well suited to the subject. How has he achieved this effect?

19

A Digression on the Complement

(a) The centre forward kicked the ball.
Kicked what?—the ball.

(b) The centre forward was jubilant.
Was what?—jubilant.

Traditionally, in sentence (a) the question 'What?' leads us to the *object* of the sentence; in (b), to the *complement*. However, most linguists now think the distinction is not worth making in English. Let us consider why the distinction between complement and object has been made in the past and whether such a distinction is necessary.

A transitive verb has been defined (*The Shorter Oxford English Dictionary*) as one 'expressing an action which passes over to an object; taking a direct object to complete the sense'. This definition seems to depend on the ancient view of cause and effect. An action was thought to 'pass over' in much the same way that the imprint of a seal was made in wax: though the original and its impression were distinct, the pattern had been transferred; was common to both. We might wonder about this a little: a cold in the head does not much resemble a draught—is this too an action passing over to an object?

Contrariwise, in sentences, such as 'Nelson was a great sailor' or, 'The road seems to be (a) clear' (road), the verb was intransitive; was thought of as a 'copula', a connecting link between two aspects of the same thing—notice how bracketed words are sometimes added to twist sentences into the required form.

The argument is that the accusative case is needed after a transitive verb to express passivity; a nominative case after an intransitive verb to express identity (through the same case) with the subject of the sentence.

The first point to be made against all this is, that even if the distinction *is* a valid one we can very rarely make it. Consider:

Peter has a dog.
A dog must have a licence.
Jip is a dog.

In the first sentence, 'dog' is the object of the sentence; in the second, it is the subject—a question of position, not of word-ending. In the third sentence, there is nothing we can do to label 'dog' to indicate common identity with Jip—but then the sentence as a whole makes that point. Does it help to call the word a complement?

In practice, only one construction in the whole of the language ever causes difficulty in this respect: a sentence which opens with 'it is/was' and concludes with a pronoun. 'It is I/me, he/him, they/them, we/us' etc. Yet the sense of the words is such that the construction can only be used in conversation. 'It is I, etc.' could not easily form part of a narrative. In conversation, usage and convention rather than logic are the rule. In fact, 'it is' will nearly always be spoken as 'it's'—and, after this shortened form, 'me' inevitably follows.

Even in logic the distinction which might be made is blurred. There *is* a tendency, which it is as well to be able to recognise, for intransitive verbs to form part of equation-like sentences—sentences of the 'oculist is an eye-doctor'-type, which tell us nothing new, but merely restate. However, what about such sentences as 'John is ready'? Is readiness a necessary result of being John?, i.e. the sentence does tell us something new. And what of 'A tree has leaves', where 'has' is a transitive verb and 'leaves' nominally the object? What action is 'passed over' here? Is having leaves (or the equivalent) not a necessary part of being a tree? In other words, the statement adds nothing to our knowledge.

Thus, I conclude that a sentence which has an 'object' may give us no new information; one which has a nominal complement may do more than merely analyse. Furthermore, 'cases' exist in English only as pronouns, and labelling nouns 'nominative' or 'accusative' is a waste of time (see chapter 25). Moreover, in practice, there is no instance of a distinctive complement form in the language.

Therefore, from this point in the book, any word that answers 'what?' after the verb, which 'completes' the verb, I shall call the complement.

A 1 Explain the traditional distinction between object and complement.

A 2 Write down all the 'intransitive' verbs you can think of.

A 3 What seems to be the logical justification for making the transitive/intransitive distinction?

A 4 How far does it hold good?

A 5 To what extent can the complement/object distinction be made in English? Answer fully.

B 1 Make up sentences in which a traditionally intransitive verb does more than act as an equals sign, and gives us new information.

B 2 Now list sentences which are tautologies (i.e. merely analyse) but have a transitive verb.

B 3 Can you think of any grammatical or logical terms which you think could be usefully abandoned or ignored—what about 'predicate' and 'indirect object', for instance? But see the next chapter.

B 4 Why on my own terms must I be careful not to assert that it is *wrong* to say 'It is I'?

B 5 English tenses can usefully be labelled simple, continuous or emphatic. Check that you know what is meant by these terms. Would it be sensible to insist that, say, the French labelled their verbs in the same way?

C 1 Here is the 'history' of 'It's me':

(a) Ic hit eom (I it am)—Old English.
(b) It am I—Chaucer's English.
(c) It is I—fifteenth century onwards.
(d) It is me—sixteenth century onwards.

Account for the change to forms (b), (c) and (d) in terms of word-order and agreement. Which of the forms would you uphold on *logical* grounds?

C 2 Discuss and illustrate the following point as made by Professor Randolph Quirk in his book, *The Use of English*: 'English has a very definite and complex grammar with some variation (such as we have in pronunciation and vocabulary) according to the social groups in which we mix, and with a small area of what we may fairly call *free variation*, where opinion may differ as to the preferable form. In examining and discussing our language, we need to observe carefully *all* the distinctions but *only* the distinctions which are made in our own grammar, and to be on our guard against phantom distinctions that are "wished on" English merely because they exist in the grammar of other languages.'

20

More Like an Equation

Stephen Leacock's 'nonsense novel', '*Q.'—A Psychic Pstory of the Psupernatural*, contains the following conversation:

'Well,' he continued, 'Q went to his office each day after that with absolute regularity. As far as I can gather there was nothing either in his surroundings or his conduct to indicate that any peculiar fate was impending over him. He saw Miss M regularly, and the time fixed for their marriage drew nearer every day.'

'Each day?' I repeated in astonishment.

'Yes', said Annerly, 'every day. For some time before his marriage I saw but little of him. But two weeks before the event was due to happen, I passed Q one day in the street. He seemed for a moment about to stop, then he raised his hat, smiled and passed on.'

'One moment,' I said, 'if you will allow me a question that seems of importance—did he pass on and then smile and raise his hat, or did he smile into his hat, raise it, and then pass on afterwards?'

'Your question is quite justified,' said Annerly, 'though I think I can answer with perfect accuracy that he first smiled, then stopped smiling and raised his hat, and then stopped raising his hat and passed on.'

This question-and-answer routine is meant to poke fun at scientific exactness—but, if we take it seriously for a moment, it seems clear that Annerly could not be perfectly certain that the order he assigned to events was the true one. However, as soon as we attempt to describe an experience we have to arrange the elements of that experience in some sort of sequence—a necessary consequence of using words.

If we were drawing a picture, we might show Q's smile and his raised hat as he is passing, but even here (unless we go beyond the normal conventions of a picture) we appear to be assigning an order to the events: namely, that they happened simultaneously.

Nevertheless, while one may agree that words are extended in time or space, to think of words as, say, a regular succession of posts which form a fence: $1+1+1+1+1+1+1+1+1$, is not particularly helpful. It is true that words may sometimes be related to events in a $1:1$ relationship; but only to achieve some special effect, as in the opening of T. S. Eliot's *Triumphal March*:

Stone, bronze, stone, steel, stone, oakleaves, horses' heels
Over the paving.

But, soon, to make sense, it has to break down: 'horses' heels' and 'over the paving' are different in character from what has gone before—not single words, but groups of words, matched to the event.

Again, if we reconsider Annerly's answer, we see that information comes less in grains than in lumps. In the same way that words vary in size and complexity, so will the groups of words which naturally go together:

{I think [I can answer (with perfect accuracy) ⟨that he (first) smiled⟩].}

S	V	C1	A1	C2	A2
u	e	o	d	o	d
b	r	m	v	m	v
j	b	p	e	p	e
e		l	r	l	r
c		e	b	e	b
t		m		m	
		e		e	
		n		n	
		t		t	

Sentence structure, then, is more like an equation than a procession.

When we read, we instinctively group the right words together. Indeed, the exception proves the rule—should we make a mistake we quickly realise that we have gone wrong. We instinctively make subtle variations in our speech—when, for instance, we use the same words twice with a different meaning. If analysis of language is to be of any use, it must be able to give an account of such subtleties: specially written exercises are easy meat.

To a large extent this is possible. The logician can explain the surprise element in many an old 'catch', and why we accept it:

Question, 'Why do white sheep eat more than black ones?'
Answer, 'Because there are more of them.'

The explanation would probably take the line that the word 'sheep' is being used *collectively* instead of in the expected *distributive* form—'fifth form' is collective, 'fifth former', distributive.

And the grammarian can explain, for instance, how the following 'chestnut' makes its point:

A: I saw a conjurer turn a bowl of goldfish into a rabbit.
B: That's nothing, I saw a man turn a horse into a field.

In A, 'into' is doing the work of an adverb, while in B it is a preposition.

One could bring out the point with brackets:

A = {[turn (a bowl of goldfish) into] a rabbit}
B = {turn [a horse] [into a field]}

A 1 What does the title of Leacock's nonsense novel suggest to us about its theme and the way in which it will be treated?

A 2 Explain, in your own words, why giving an account of any happening involves assigning an order to events.

A 3 Why is it not very helpful to think of words as being like posts in a fence?

A 4 In what ways does sentence structure seem to be like an equation?

A 5 Using your own words rephrase the question 'Why do white sheep eat more than black ones?', making it clear that 'sheep' is being used collectively, and ruling out possible catch-answers.

B 1 Using brackets, group the words which hang together (as in the boast about the conjurer) in the sentences below, and then decide which of the 'direction' words are prepositions, and which are adverbs:

Turn down the road and you'll see it.
But, please don't turn down the corners of pages.
I ran down the lane.
She runs down her neighbours.
He fell in the river.
I fell in with the scheme.
Can you come in the other room for a minute?
We'll come in the car, then.

B 2 With the aid of brackets, bring out the different senses of the following:

An old clothes shop; a dog loving man; a stainless steel sink; ten pint bottles; the daughter of Pharaoh's son; the missing child's hat; six foot men; a man eating fish; the Prime Minister keeps his secret still; Peter says Paul is a liar.

B 3 (a) Use the word 'very' in two successive sentences, so that in the first it has its normal adverbial use (i.e. it qualifies an adjective or another adverb), but in the second is an adjective. (b) What part of speech is the word 'a' in 'once a day'?—careful!

B 4 'Specially written exercises are easy meat.' Is there, then, nothing to be said for the text-book? Take any educational book you wish, this one if you like, and analyse it on the following lines: How much of it is (a) designed to fix facts in your memory, (b) to establish a principle, (c) to make you think for yourself, (d) to entertain you, (e) to widen your interests, (f) merely to keep you quietly, but pointlessly, at work? What conclusions do you come to?

B 5 What do you understand by: grammar, logic, syntax?

C With regard to the conventions of a picture, look for reproductions of the so-called 'Futurist' school—especially of 'A Lady And Her Dog' by Giacomo Bella. What is the relevance of such a picture to the present issue? What is your reaction to this and other experimental art? Do you know of any 'experimental' writers or musicians? Is the word 'experimental' justified in this context? What has been achieved by such 'experiments'?—or are we wrong to look for a 'success' that can be measured?

21

The Basic Sentence Analysed

The centre forward / booted / the ball / into the top of the net.
subject verb complement adverb

But suppose that the sentence had taken the following form:

The centre forward / who was dropped last week / booted
subject adjective verb

the ball / into the top of the net.
complement adverb

So far so good; yet isn't 'was dropped' just as much a verb as 'booted'? If we treat the one as a separate entity, must we not do the same for the other?

The centre forward / who / was dropped / last week
subject subject? verb adverb

booted / the ball / into the top of the net.
verb complement adverb

Such an analysis is merely muddling—the structure of the sentence is no longer clear and there is the feeling of 'spare parts' about the whole thing.

What, for instance, is 'who'? Certainly we could label each word in the sentence:

The / centre / forward / who / was / dropped / last
definite article adjective noun pronoun verb participle adjective

week / booted / the / ball / into / the / top
noun *verb* *definite article* *noun* *preposition* *definite article* *noun*

of / the / net.
preposition *definite article* *noun*

Such a list of parts may be useful in making comparisons from one sentence to another, but as regards this sentence itself it tells us nothing: seventeen existing words have merely been related to twenty-one new ones. We have gathered nothing of the architecture of the sentence. It is as if we had wondered how a building managed to keep standing and had been told: 'stone, wood, mortar, plaster, nails, steel'; instead of an answer in terms of strains and stresses, weights and balance.

Things might become a bit clearer if—to keep the image of a building—we distinguished between different levels of meaning. Thus at the one level 'who was dropped last week' is an adjective; at the next, it has the structure 'pronoun verb adverb'; at the lowest, pronoun, verb, participle, adjective, noun.

Let us consider the two simple sentences which make up 'The centre forward who was dropped last week booted the ball into the top of the net'. They are:

(i) The centre forward booted the ball into the net;
(ii) The centre forward was dropped last week.

Sentence (i) has the form SVCA; sentence (ii) SVA. S is the pivot:

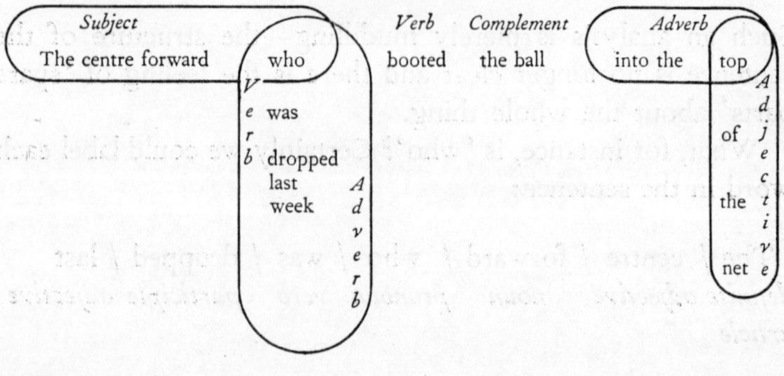

Sentence (ii) functions as an ADJECTIVE in the whole.

Now, the 'basic' sentence in its simplest terms might take the form:

subject	*verb*	*complement*	*adverb*
He	kicked	it	there

Retaining the existing pattern, the complexity of the sentence might be increased (a) by using phrases for each section, (b) by expanding the phrases into clauses.

So the subject might become:

(a) The stand-in centre forward; or

(b) The centre forward who had been dropped last week.

The two sentences as a whole might take the form:

(a) The stand-in centre forward managed to kick the rain-soaked ball into the top of the net; or

(b) The centre forward who had been dropped last week managed to do what the crowd wanted when he kicked the ball into the top of the net.

Structure becomes simplified as it moves upwards (see diagram on p. 80) or—to put it another way—particular details which are essential to the structure are lost in the general pattern as the mind's eye withdraws. If, perhaps, we were looking very closely at a chain-link fence we might be preoccupied with the bending and linking wire; at a little distance we should see rather the honeycomb or diamond pattern which the wire formed; further away still we should be aware only of the horizontal grey outline of the fence itself.

A 1 Why does the form of analysis used in the previous section prove inadequate? What new concept is introduced which enables it still to work?

A 2 SVCA analysis might be objected to as illogical, though I have used it because its terms are familiar. Can you spot the difference in nature between the labels 'subject' and 'complement' on the one hand and 'verb' and 'adverb' on the other?

'Layers' of structure—how the sentence 'stays up'

(Though it came as a surprise to me, the similarity of this diagram to the outlines of a marquee is not insignificant)

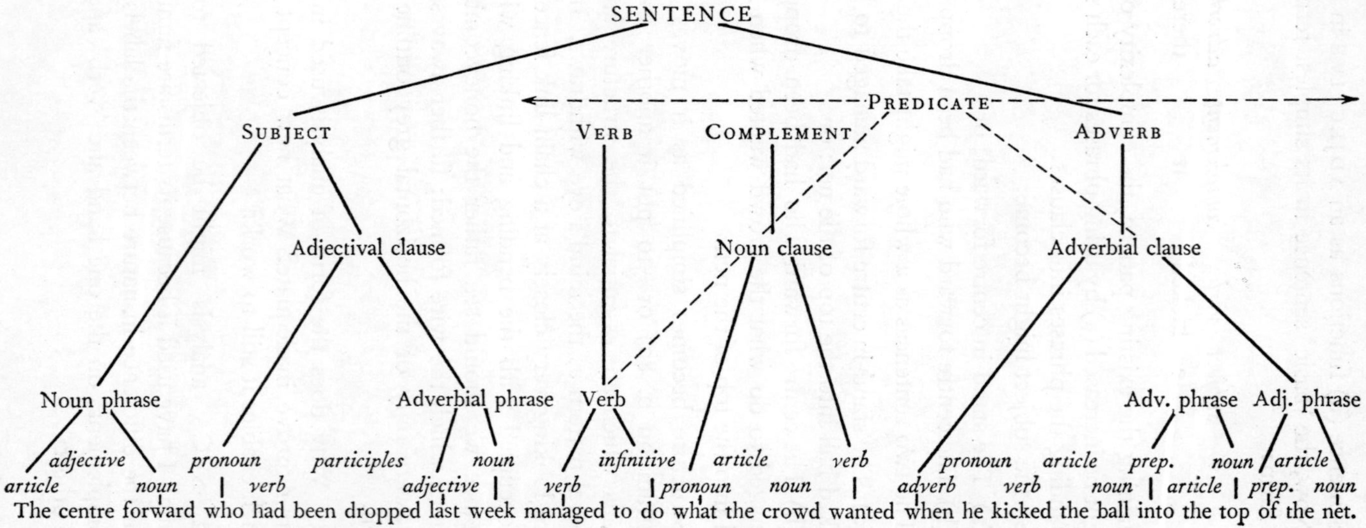

A 3 In the two simple sentences (i) and (ii), why has (ii) no complement?—or, to put the question another way, what is the logical subject (as distinct from the grammatical one) in this case?

A 4 My examples are the best I can think of, but not always ideal. For instance, having shown the subject of the basic sentence as (a) phrase, (b) clause, why did I not go on to do the same with the complement?

A 5 Taking 'He left the house then' as a basic sentence, expand SVCA to phrases, SCA to clauses. Then label each part and plot the structure of the sentence (as in the diagram).

B 1 A justification of analysis would, I suppose, take the following lines: (i) it clarifies meaning, drawing attention to the kind of information which each part of the sentence contains; (ii) language is a human artifact and as such should be studied as carefully as any archaeological find—more so, as the surest way to examine the human mind is to examine the results of it at work: and language is the most complicated, distinctive and accessible record of this. What do you think of these arguments? Can you think of other lines of defence, or attack?

B 2 Which seems to you the most worthwhile level at which to analyse—that of word, phrase or clause? For instance, 'who had been dropped last week' is in its context an adjectival clause, yet the emphasis is on *time*, not *kind*.

B 3 Subject and predicate have commonly been the theme of many exercises. Can you see any benefit in introducing the term 'predicate' which I have added to the structure diagram?

B 4 Find out what you can about the term 'predicate' and, if possible, the varying way in which it has been used by grammarians and logicians.

B 5 In the structure diagram, the main clause is 'the centre forward managed to do'. But clause analysis remains examination fodder for the most part. To what extent do modern authors write in terms of clauses? Consider any page of recent writing: what proportion of sentences consists of main and subordinate clauses? When there are main clauses, how often are they connected not to subordinates, but to co-ordinate clauses? Clauses, that is, of equal status to the main clause and linked with it by such words as 'and', 'but', 'so'. How much of the writing is analysable only at phrase level?

C Apropos the notion of structure: perhaps because of the make-up of the human mind, we find this diminishing/expanding pattern everywhere. Earth goes round sun, moon goes round earth, electron goes round its nucleus. The structure can be seen in architecture, for instance, partly because architecture is a product of the human mind, but also (though there might be other ways of achieving this end) so as to keep the building up. Look at the history of, say, Gothic architecture in terms of weights and stresses. Ask yourself such questions as, why did buttresses grow deeper as windows got larger? What are flying buttresses and pinnacles? What purpose do they serve? Are gargoyles merely funny faces? Carving, merely decoration? Note that words which are normally applied to language are sometimes applied to the arts—the phrase 'the Gothic *idiom*', for instance. What has given rise to such comparisons?

22

A Flexible Language:

The English language has always been, is still, in the process of changing. It has even been suggested that the language of the future will be developed not from B.B.C. English but from that spoken in the West Indies, say, or in Australia. This idea has been put forward because the same sort of thing has happened before—long ago. Colonial Latin, spoken in the provinces of the Roman Empire—'broken', imperfect Latin, that is—grew into modern Italian, French and Spanish. In the meantime the 'correct' official Latin 'died'.

Anyway, the English language is very flexible. From at least the time of Shakespeare it has not been hampered by a lot of rules which have to be stuck to. A noun, for instance, may readily become a verb; sometimes it adds '-ise' (sterile, sterilise), sometimes it does not even change its form—the word 'question', for instance: (a) he asked me a question; (b) he questioned me. But here there is a difference of meaning—what? While you are at it, what obvious advantage has sentence (b) over sentence (a)?

As for words ending in '-ing', they can normally be used as a verbal noun (subject or complement of a sentence) as an adjective, or as a verb (in combination with 'am', 'is', 'are', 'was' or 'were'). For example:

> *Gardening* is a good way of getting blisters.
> I was *gardening* at the time the wall fell down,
> I had on my *gardening* boots.

Which is noun, which verb, which adjective?

Proper nouns (names of people and places) that have become applied to everyday things have already been mentioned. Here are half a dozen more. Find out the meaning, if it is not obvious, and as much as you can about the person or place from which the name comes: bedlam, spoonerism, florin, jersey, sardine, sherry.

You were also asked to make a list of words such as orange, lilac, silver, lemon, peach, gold, ruby, emerald. Words, that is, which were originally names of things (nouns) but which are now often used as colour words and so are called— ?

Then there are cases of gradual change of meaning. 'Deer', for instance (the modern German word 'Tier') once meant all animals; 'meat' meant all food; 'fowl' all birds; 'hound' all dogs. Such words as these have become much more precise in meaning. On the other hand terms such as 'education' (which originally meant 'drawing forth') have become much more abstract and vague. The word 'consider' ('to look at a star') is another case which was mentioned earlier. List ten such words which mean much the same as 'think'. How many of these verbs for mental processes appear to have been used in the first place about things in the outside world? (They are 'dead' metaphors, that is to say). 'Consider' itself is one example.

A 1 What do you understand by (a) the provinces, (b) colonial English, (c) broken English, (d) B.B.C. English?

A 2 Why do you think single inverted commas are used in paragraph 1 for 'broken', 'correct' and 'died'?

A 3 What suggestion is made about the languages of the future?

A 4 What evidence is given for the statement that our language is a flexible one?

A 5 What is meant by saying that some words have become more abstract?

B 1 Guard, part, shadow, house, sight, mount, close, ride, record, fence, sky, refuse, desert, envelope, supply, report.

Use each of the words in sentences (a) as a noun, (b) as a verb. Which of the nouns change their pronunciation on becoming verbs?

B 2 Make up a rule which will enable you to use the words 'practice' and 'practise' correctly. Both are pronounced the same, unlike 'advice' and 'advise', which they otherwise exactly resemble in 'behaviour'.

B 3 Make a list of verbs ending in '-ise'.

B 4 Sir Ernest Gowers in his *Complete Plain Words* notes that the following nouns are tending to become used as verbs: contact, feature, glimpse, position, sense, gift, author, underground. Which of them, would you say, are in general use as verbs?

B 5 In the same general way as the 'gardening' example in paragraph 3, use each of the following words as (a) verb, (b) adjective, (c) noun: opening, working, lining, feeding, fencing, questioning, facing, waiting, painting, cooking.

C 1 Another look at the sentence 'the dog bit the man' may bring out the point about flexibility. The sentence contains two nouns—both of which can act as verbs by changing their sentence position from first or third to second and adopting the appropriate endings to indicate tense. In the German equivalent (Das Hund biss den Mann) this procedure is not so simple! 'Der Mann' with the addition of prefix and suffix can become the verb 'bemannen', but, though we have the verb 'to hound', 'hunden' in German does not exist. 'To dog' is rendered 'nächspuren' (compare our word 'spoor'), to 'go after the track'. This comparison makes two points about language: (a) that the English *content words* (see chapter 28) are interchangeable to a high degree; (b) that English makes compound words much less readily than German or Greek, or, for that matter, Old English.

As a further example, take the names for the principal parts of the body, beginning, perhaps, with 'eye', 'mouth' and 'head'. How many of these nouns can function also as verbs? How many of the German equivalents form even the basis of verbs?

C 2 Verbs too can become nouns. Using *The Shorter Oxford Dictionary*, select from those words which can be used as both noun or verb, (a) ten examples which were used first as noun, (b) ten examples which were used first as verb. Note the date when the word's usage widened. Do you find that a large proportion of the examples were first used in both ways in the sixteenth century? Can you explain why this should be so?

23

. . . which is still growing: Coinage and Onomatopoeia

As long as we stick to our normal word-order, even nonsense can sound convincing:

> 'Twas brillig and the slithy toves
> Did gyre and gimble in the wabe:
> All mimsy were the borogoves,
> And the mome raths outgrabe.

That is the first stanza of the 'hard' poem with which Alice was confronted in 'Looking Glass Land'. Although she did not understand it, Alice was convinced that it had a meaning. Convinced, I think, because of the normal-sounding word-order used (the sections in italics are wholly or in part 'nonsense'):

> Subject ('it') verb *noun* conjunction *adjective noun*;
> *Verb* conjunction *verb adverb phrase*;
> Adverb *adjective verb subject*;
> Conjunction *subject verb*.

However, one could take the final sentence to read:

> And the mome raths outgrabe
> *noun verb noun*
> (Compare 'mom likes grapefruit'.)

It all depends on whether one takes 'out' to be a real word or nonsense.

To some extent, then, it is the word-endings which help the illusion—'raths' can be seen as either plural noun or singular verb, there is no temptation to think it might be, say, an adverb or a singular noun. Which other endings in the verse are characteristic of which parts of speech? Which word sounds (from the point of view of ending) least like English? If one takes 'out' to be a real word, is the structure of the sentence entirely English in character?

The grammar of the sentence is orthodox. Try putting the poem into the present tense. The fixed pivots of 'it', 'and' and 'the' are real enough, and the word-endings right; but the vocabulary is capricious (see chapter 28).

In the book, Humpty Dumpty, an odd logician, explained some of the words. Here is what he had to say about one of them:

'Well, "slithy" means "lithe" and "slimy". "Lithe" is the same as "active". You see it's like a portmanteau—there are two meanings packed up in one word.'

One may coin and define as many words as one wishes, but it is usage, not a wave of the magic wand, that gives them meaning. Nevertheless, certain of Lewis Carroll's word-inventions have passed into general use. 'Chortle', for instance, is given in dictionaries as meaning a low deep laugh; so is 'burble'—to talk incoherently.

Both these expressions were originally 'portmanteau' words; but they could also be considered as examples of onomatopoeia—of the same order as: 'whistle', 'bang', 'twitter' and 'growl'. 'Said', 'called' or 'asked' are also to do with sound—what is the difference between these two lists?

Once a catchy name has been found for a thing (or an idea) it becomes accepted—this is the notion behind brand-names and trade-marks. Brand-names are proper nouns and, as such, outside the scope of a dictionary. However, one or two trade-names have become so successful as to become associated with all articles of a certain type, but no longer with a particular firm: you will quite often see in advertising frames the inscription 'hoover wanted—any make'. This is true of other products: if you went into a chemist's and asked for a jar of petroleum jelly, you might very likely not be understood; the brand-name has become so well known as to cease to be effective publicity for the manufacturers.

Trade-names are not our concern here, though some of them seem to be passing into the language as common nouns. But a similar and increasingly popular way of devising a catchy name has already provided dictionaries with a few entries. The subjects are usually organisations or pieces of apparatus. Here is a list of such words, some of them accepted as permanent, others only passing: radar, posh, 'wrens', Reme, Ernie, Uno, Unesco, Unicef, Nato, Aslef. What have these 'words' in common? Find out their origin.

A 1 Pick out the genuine words from the nonsense verses quoted.

A 2 If 'chortle' and 'burble' began as portmanteau words, can you suggest which other words they come from?

A 3 Why 'portmanteau' words?—what is a portmanteau?

A 4 'Some brand-names have become almost too well known.' Explain what is meant by such a comment.

A 5 In what way are words such as 'Ernie' rather like brand-names?

B 1 Supply the missing verbs in these sentences: Donkeys —; turkeys —; hens —; horses —; cats —, or —, or —; dogs — a warning, or — at strangers, or —, if hurt, or — when puppies.

> I'd rather be a cock and —,
> But I dunno.

B 2 Pop, clatter, crash, tinkle, bang, splash, plash, clang, boom, rattle, ring, splosh, crunch, spludge, lap... are all words which imitate sound. Which sound wet? Which metallic? Which have a comic effect? Which are pleasant to hear? Which unpleasant? Which neither? Which is the loudest? the softest?

C 1 Alice said that the poem filled her head with ideas, only she did not know exactly what they were. Read the whole poem—what do you think? Also attempt explanations of the nonsense words in the stanza quoted—'wabe' and 'mimsy' are clearly portmanteau words—and compare your 'meanings' with those given in chapter VI of *Through The Looking Glass.* Look at the preface to that book; opposite it there is a diagram—explain its purpose and the way it works.

C 2 It is usage that gives a word meaning.... The words 'zog', 'squatchit', 'grelf', 'mhall' and their derivatives have, for the sake of argument, been recorded in the contexts given below. Write as complete an account as you can devise of the 'meaning' of each word, distinguishing between literal and figurative use.

I was near Zog bridge.
A zog is tastier with sugar.
The original zog is no longer made.
Zog Night dates back to medieval times.

The squatchit net was burst by the hurled hoop.
The hoop ringed his squatchit stick precisely.
The whistle went for quarter time in the squatchit match.
The twenty seven squatchit players left the field.

Grelfully, I clasped her hand.
A grelf was growing by the wayside.
A bottle of Grelf Number Seven, please.
Keep your skin grelful.

I feel mhalled today.
The 'cello mhalled in the background.
Mhallfully, I make my way.

24

Pronunciation and Spelling

A famous fictitious Frenchman, after mastering most of the difficulties of English spelling, was finally convinced that 'ough' (as in 'lough') represented the sound 'ock'. Once more, and fatally, was he corrected: he stuck to his guns and laid his instructor low 'with a rough'. Another student is said to have shot himself on seeing a newspaper report headed '"Hamlet", Pronounced Success'. On a different level of eccentricity, it has been pointed out that an alternative spelling of the word 'fish' is 'ghoti'. The alternative is acceptable if 'gh' is sounded as it is in 'cough', 'o' as in 'women', 'ti' as in 'nation'.

In Chaucer's time, English spelling was more nearly phonetic. If you have studied German you will know that 'dictation' in that language is easier than it is in French or (even for us?) in English—because once certain rules are learnt, one knows that the language will keep to them. In German, 'ei', for instance, will always be pronounced as in the word 'eiderdown'. To demonstrate the caprice of our own language, one has only to consider the words 'their', 'either' and 'perceive'. Fortunately, our consonants can, on the whole, be depended on.

Historically too, they are more stable: their basic form had been settled a thousand years before Chaucer. Indeed, the Germanic consonants are as old as the Germanic language. Slowly a different pattern of sound had emerged—at the point where the pattern ceased to change, the Germanic language became distinct from the rest of the Indo-European group (see chapter 10).

Two factors govern our articulation of a consonant—the extent to which we interrupt the breath passing through our mouth and nose; and the degree of 'voice', the extent to which the vocal chords are allowed to vibrate. It has been suggested that there may have been a physiological basis to the Germanic consonant-shift.

It was Jacob Grimm (joint editor of the famous *Tales*) who first published the numerous instances from which general hypotheses (generally known as 'Grimm's Law') could be induced. For example, our verbs 'be', 'do' and 'go' can be traced back to the oldest written language of the group, Sanskrit—as 'bhu', 'dha' and 'ha'. The Indo-European consonants, bh, dh, gh, lost their aspirates to become b-, d-, g-. 'Brother' and 'daughter' are two other words which can be traced back in a similar way.

The other main changes postulated by Grimm's Law can be shown conveniently through Latin. It should be understood, though, that the Old English words are not derived from Latin: both Latin and Old English have a common Indo-European 'ancestry', but Latin has preserved the consonants in their old form. (See page 93.)

Chaucer's 'Middle English' was still near to German in much of its pronunciation, and certainly like it in its consistency. In present-day English final e's, for instance, seldom have a sound of their own; though they do serve to modify the central vowel of the word or syllable. (Thus, we do not need tiresome accents and umlauts—compare 'lath' and 'lathe', 'Rad' and 'Räder'.) 'Name' in fourteenth-century English, as in modern German, was Nah-me. In the next hundred years the sound seems to have shortened to 'nam'. By Shakespeare's time it had lengthened to 'nerm'; in the next century become 'pure' as in the French né. From the eighteenth century on it seems to have assumed its modern diphthong form (é + short i). Yet, throughout, the spelling stayed the same.

This catalogue of changing sound has a pattern. Pronounce 'name' in each of the ways mentioned in the paragraph above. In each case try to locate the point on the soft palate where the tongue and lips seem to focus the sound. Working from the old *ah* form, what do you notice about the 'position' of successive sounds? The 'movement' of sound which you should have observed is general—it does not apply merely to the present case—and has become known as the Great Vowel Shift.

	Examples		
Nature of change	*Latin*	*Old English*	*English*
(1) Aspirate added:			
p becomes f	piscis	fisc	fish
	ped-	fōt	foot
t becomes th	tenuis	thynne	thin
	tu	thu	thou
	tres	thrī	three
k becomes h	centum	hund-	hundred
	cornu	horn	horn
(2) 'Voice' lost:			
b becomes p	tribus	throp	thorp (village)
d becomes t	domare	tam	tame
	dent	tōth	tooth
	decem	tien	ten
g becomes k	genu	cnēo	knee
	genus	cyn	kin

A 1 Explain the misconception about 'Hamlet'.

A 2 Why should German be easier to spell than French or English?

A 3 The Germanic consonants are as old as the Germanic language—explain.

A 4 Give examples of voiced consonants.

A 5 Why should Latin be referred to in the chart if the English words are not descended from the Latin ones?

A 6 What is Grimm's Law and why is it so-called?

A 7 Compare the function of the final 'e' in modern English with that of the umlaut in German.

A 8 Give as full an account as you can of the Great Vowel Shift.

A 9 Why do you suppose spelling remained unchanged as the vowel-sounds altered? Compare with the case of the consonants.

A 10 Find other 'alternative' spellings for common words.

B 1 Explore the 'ough' territory for yourself. Write up your findings in the form of a report or the story of a misunderstanding.

B 2 The many foreign words in the language have given English a great variety of synonyms, so that many shades of meaning can be expressed. In a sense, though, this influx may have weakened the language: it has meant that certain nouns, adjectives and abstract nouns which are connected in meaning seem to have no formal or linguistic connection. Here are half a dozen examples: brother, fraternity; father, paternal; mother, maternal; son, filial; nose, nasal; eye, ocular. However, it may be possible to establish an indirect formal relationship. Using a good dictionary, common sense and Grimm's Law where applicable, try to determine which of the above pairs of words can be traced to a common 'ancestor'. Having done that, attempt the examples below:

heathen, pagan; sitting, sedentary; tooth, dental; both, pair; mind, mental; kneel, genuflect; tame, domestic; tenuous, thin; hostage, guest; warden, guardian; double, biennial; answer, respond; kind, gender; eight, octopus; captain, head; wordy, verbal; tongue, language; red, ruby.

B 3 The following amorous lines from Shakespeare's *Venus and Adonis* are interesting not only for their content:

> But O, what banquet wert thou to the taste,
> Being nurse and feeder of the other four!
> Would they not wish the feast might ever last
> And bid suspicion double-lock the door?
> Lest jealousy, that sour unwelcome guest,
> Should, by his stealing in, disturb the feast.

('The other four' refers to the senses.) What comment have you on the rhymes?—it is unlikely that they were simply 'bad'. Try to determine Shakespeare's pronunciation of the sounds in question by browsing about his works. As far as you can tell, would 'four' rhyme with modern 'door' or with 'flower?'—and what of 'taste' and 'last', 'guest' and 'feast' (fest?)? Or, did the four words have the same vowel-sound? Check the rhyme scheme with the rest of the poem.

C 1 Consider the word 'take' as spoken in Scots, Lancashire, Black Country and Cockney dialects. What conclusion do you come to about the relationship of (a) sound to spelling? (b) existing speech to the history of speech?

C 2 The English dialects (regional pronunciation and local vocabulary) are tending to disappear, it is said. Give some reasons. Are there any phrases, not now in general use, which you have heard from old people? For instance, my grandmother would refer to the 'horse-road' and the 'ashpit'.

C 3 Recent 'research' at Sydney University has established that the orthography of *Strine* is distinct from that of English. Here are some examples:

Tiger: An imperative, as in 'Tiger look at this'.

Inner narkup ludaze: A term from the building industry, 'I'll have her finished inner narkup ludaze'.

Gloria sarah titch: Phrase used by Strine statesmen, 'We honour our gloria sarah titch with our last droppa blood'.

Split nair dyke: A medical term, 'I gotta split nair dyke'.

Gissa lenya: An informal request, as in 'Gissa lenya coam.'

See Shaw's *Pygmalion* for an attempt to set down Cockney. If you have a good ear for speech-sounds, try a similar transcription of dialect or (and this is probably easier) of a foreign language.

C 4 It has been suggested that school children are the last tribe to defy the effects of civilisation. Certainly they are a good example of a community within a community: schoolboy speech is long-lived and flourishing. For instance, at least one of the words used to gain a respite in a game goes back to Chaucer's time. Which was the 'magic' word in your young days? Is there a variety of these truce terms within your group? Do terms seem to relate to locality? How are they distributed geographically? Can you trace more than half a dozen?

Make a collection of characteristic comic suffixes used in child language; for example, -gog, as in 'goosegogs'. List the traditional nicknames and epithets which go with certain names and physical characteristics. What are the time-honoured substitute-names for various kinds of food? for the subjects of the school curriculum?

If you have gone this far and are still interested, you will be able to continue under your own steam. You might get hold of *The Lore and Language of School Children* by Iona and Peter Opie, and see whether your own researches can augment or correct the mass of information in that excellent book.

25
Some Words Still Vary with Their Job

To come back to the 'understandable' German sentence: *Das Hund biss den Mann.* So firm is the grip of the normal English sentence-order on us that we tend to think that the first unit of the sentence must be the subject, the second the verb and the third the complement—as, indeed, is the case in:

Der Hund biss den Mann.
The dog bites the man.

Yet, this sentence:

Den Mann biss der Hund
— man bites — dog } means 'the dog bites the man',

as surely as the previous one; though someone who was entirely ignorant of German grammar but had a German–English dictionary would most likely translate it according to the word-order as: 'the man bites the dog'.

However, the next sentence, *Der Mann biss den Hund*, and this one, *Den Hund biss der Mann* both mean 'the man bites the dog'.

In German and Latin and Old English most words alter according to their function in the sentence. The German writer, for instance, does not have one 'cut and dried' word for 'the', but has the choice of: *der* Mann (subject of sentence), *den* Mann (object of sentence: the distinction between object and complement is necessary in German); *des* Mannes (belonging to the man—the man's) and *die* Männer (plural—pronounced like 'men').

In a language which has many of these changes *word-order* is not so important—you tell whether it is the man or the dog that is doing the biting not by the order, but by the form the word takes. Old English in its word-behaviour was much like modern German—only more complicated still. (In the extract which follows the letter 'thorn'—an unnecessary difficulty—has been replaced by 'th'.)

'Eft tha Gregorius befran, hwaether thaes landes folc cristen waere othe haethen, him man saede thaet hi haethen waeron.'

(Again when Gregory asked whether the folk of the land were christian or heathen, someone said to him that they were heathen.)

'Waer*e*' and 'waer*on*' are both translated by the modern 'were'; 'saed*e*' is 'said'; 'haethen*e*' is 'heathen'. On the whole, modern English is simpler: a lot of lopping of ends has occurred.

Yet not all of them are gone.

What is another way of saying 'the folk of the land'? Or, more naturally, an alternative for 'the hat of the lady'? What of 'thaes landes folc', then? And what of 'him'? 'Him' and 'he' may be the same person, yet they are different words. The job done in the sentence still has some importance, in spite of the emphasis on word-order.

The most notable instance of word-change is the addition of the apostrophe 's' to show 'belonging to'. You may have been told in the past that an apostrophe either means a letter has been missed out or indicates 'possession': we can now replace this with a simpler rule—an apostrophe always means a letter has been missed out, usually an 'e' or an 'o'.

But as our language has comparatively few of these word-changes, the clearest sort of English sentence is that in which subject comes first, verb second, and so on. There is no rigid rule that says the subject must come first, but we take a lot of convincing that the beginning of the sentence is not its subject.

'The man bites the dog'

can only mean, unlikely as it is, that the dog was bitten. In a language such as Old English or modern German our first thought would probably be that a slight mistake had been made in the form of the words and the dog was really meant as the biter. Certainly, in these languages, if we said 'me bit the dog', it would be quite clear that the dog had bitten us. But, in English, so firm is the conviction that the subject comes first in the sentence that 'me bit the dog' would most likely be taken as an ungrammatical version of 'I bit the dog' rather than an un-English version of the more probable 'the dog bit me'.

A 1 How is it that *den Mann biss der Hund* and *der Hund biss den Mann* both mean 'the dog bites the man'?

A 2 Explain what is meant by saying 'him' and 'he' may be the same person, though they are different words.

A 3 Why is word-order so much more important in a simple sentence in English than it is in German?

A 4 How many different forms of 'der Mann' are there in German? In what ways can 'the man' alter in English?

A 5 Explain the 'simpler rule' about the apostrophe.

B 1 Why is there no apostrophe in 'Yours faithfully'?

B 2 The notion of possession implied by an apostrophe is a very broad one. Analyse the form of usage in each of the following examples. If not 'possession', what could each relationship be called?

The boys' overcoats; five o'clock; the Duke of Bedford's estate; a week's holiday; a stone's throw; Peter's departure; the boy-I-told-you-about-before's idea; a friend of my brother's; England's team; William and Mary's reign.

B 3 In the light of the above examples, what do you think is meant by (a) a double genitive, (b) a group genitive?

B 4 Given a compound subject (as, for example, the third in the list above) how does one (a) indicate possession, (b) form the plural? Write down a rule to cover this usage.

B 5 An alternative to the -'s case ending has been used in English since the twelfth century—the French construction involving 'of'. In each of the following instances, say which of the methods of forming the genitive is most appropriate and why. If either could be used, do you detect any variation of meaning or emphasis because of the form chosen? For example, given the two nouns 'mayor' and 'town', one could say either 'mayor of the town' or 'the town's mayor'. The former is probably the more natural, though the second version might be used if the town was the main topic under discussion.

mayor, chain; week, sick-pay; Boswell, *Life of Johnson*; children, shoes; Britain, history; Wellington, Duke; Tarzan, son; I, heart; my sister, friend; today, delivery; John Brown, body; Keats, house.

In which examples would prepositions other than 'of' make better sense?

C 1 The apostrophe is rather like the French circumflex, as in: la quête; un arrêt; un hôte; la fête; la côte; le prêtre; le maître; l'île; la forêt; les vêtements. In former times, these words contained another letter—which? Write the examples down, inserting this same 'missing' letter after the circumflex. Which English words appear to be derived from these old French ones? Compare the current French meaning with that of the related English words—which of them have changed their sense?

C 2 Analyse in terms of subject, complement, verb, and adjective/adverb-equivalent the following Miltonic variation on SVCA. Here, man comes very near to biting dog:

Him the Almighty Power
Hurld headlong flaming from th' Ethereal Skie
With hideous ruine and combustion down
To bottomless perdition, there to dwell
In Adamantine Chains and penal Fire,
Who durst defie th' Omnipotent to Arms.

26

Grammar and Logic:

Penguins do not fly.
Dragons do not exist.

These two sentences are of the same grammatical form. 'So much the worse for grammar', says the logician whose predecessors were tempted to suppose that dragons must in some sense exist—by the very fact that we are able to talk about them, to name them.

Take another pair of sentences:

It's me.
It is I.

It is commonly supposed that, though the first sentence is in wide every day use, and the second hardly ever used, it is the latter which must be correct. Similar status would be given to the question, 'Whom were you with last night?' and denied to the form which is preserved as the title of a music-hall song.

If asked why the one is right and the other wrong, you might answer:

(a) Don't know.

(b) Because one sounds right/wrong/posh/affected/pedantic, etc.

(c) It is a question of case—the verb 'to be' takes the nominative, while 'whom' is the accusative form of the relative pronoun. 'Whom' is attached grammatically to 'with'—and in English all prepositions take the accusative.

People sometimes say that they learnt their English grammar for the first time when they began to study a foreign language. Such a statement may well be intended as a rebuke to teachers, text-books and the English language itself! What can validly be claimed is that it may be a help in learning, say, German to know the function of the so-called indirect object, though as far as understanding English goes it makes more sense to think of it as an adverb.

Again, you have probably worked your way through exercises of the 'pick out the unnecessary words from the following'-type. Number one could be Shakespeare's 'most unkindest cut of all'. The 'double-negative' is another favourite for casual oral work: 'I haven't got none, Miss.' 'Ah,' says the teacher, 'in that case you have got some, for you denied that you had got none', etc.

A logical language might conveniently form its negatives, not by the clumsy device of additions such as 'do not', but by simply inverting the verb. The opposite of 'I am going' would then be 'I ɐɯ ƃoᴉuƃ'. Clearly, a double-negative would invert the already inverted verb, so turning it right side up again.

In logic, then, no double-negative is possible: but there is a difference between what ought to be and what is—a geographer, for instance, does not criticise an escarpment because its slope lacks regularity. The double-negative has an ancient lineage: 'I ne seye not' was the accepted fourteenth-century form, which later became 'I say not', finally 'I do not say'. (Apart from the spelling, the first example sounds less archaic than the second—why?) Moreover, no one is taken in by the pretence that the meaning of 'I haven't got none' is unclear. The problem is one of rhetoric, rather than of logic. The double-negative is a trick of emphasis—one which is no longer accepted by the educated. But, though we reject it, it is on grounds of usage rather than by argument. If we were to use reason to make head against the English language we must attack all our idioms for a start. Consider the uses we put even one word to, the verb 'do'—how many of them are logical?

A 1 Check that the first two sentences *are* of the same grammatical form—do the verbs differ in their nature?

A 2 Why might sentence structure have tempted us to suppose 'dragons' a reality? Take the 'penguin' sentence into account—there we do *appear* to be referring to specific creatures.

A 3 Why is it odd that 'It is I' should be considered 'correct'? (You have done this one before!)

A 4 What are the standards of 'correctness' involved in such a judgment?

A 5 What do people really mean when they claim to have learnt English grammar by studying a foreign tongue?

A 6 Rephrase economically 'most unkindest cut of all'. Defend the original version.

A 7 For practical purposes why would turning the verb upside down be 'clumsy' too?

A 8 How exact is the comparison between the role of the geographer and that of a grammarian?

A 9 Is using language in an illogical way defensible?

A 10 Compare illogical usage of the 'I haven't got none'-type with that involved in 'how do you do?'

B 1 Continue the conversation between the child and the school teacher, so that the child, while retaining his character as a child, proves of unexpected calibre and gets the better of the argument.

B 2 'She can't seem to settle down, not even after a sedative.'
'But how awful for her.'
'Yes. Even as a child, though, she had to have two night-lights and six dolls to go to bed with.'

This short conversation contains: a double-negative; a 'sentence' which opens with a conjunction, but has no verb; another which ends with a preposition. All 'errors'. However, which of the 'mistakes' are in accordance with current educated usage? Which in their present context would become pompous if re-expressed? Which sounds awkward or lacking in clarity as it stands? Do you go on to conclude that a preposition is something that one must never end a sentence with?

C The student of language is normally preoccupied with a word's connotation; its application and use. The distinctive concern of the logician is with denotation, into which class a word comes: the class of names which refers to 'real' things, or some other. To come back to the 'penguins' and 'dragons' sentences, the logician cannot admit that 'dragons', the grammatical subject of the sentence, is a name—for it would have to have an *actual* counterpart to refer to. Consequently he must also deny that 'penguins' in the previous sentence is used in a referential way. Both could be restated as definitions, making it clear that words and not entities are in question:

'Penguin' = non-flying bird (though the word might be used referentially, e.g. 'that penguin has a yellow chest').

'Dragon' = fire-breathing beast (a word that cannot be used referentially, as a name).

In such a sentence as 'Fido bit me', the grammar is not misleading: both 'Fido' and 'me' are names, they 'refer' back. On the other hand, there are a number of words which have the same grammar as names. 'Something', for instance: 'Fido bit something.'

The test for a name in logic is whether there is a criterion of identity. From the sentences,

'Fido bit me' and 'Fido is my dog'

it is apparent that Fido, my dog, bit me. On the other hand, if we take two sentences using 'something':

'Fido bit something' and 'Something woke me in the night'

it would be foolish to assert that whatever it was that Fido bit woke me. 'Something' is not referring in both cases to one particular thing.

C 1 Give an account of how we recognise (a) a noun, (b) a name in the *grammatical* sense.

C 2 Try to devise other sentences in which what looks like a name in *logic* is not, in fact, so.

C 3 Traditional logic is based upon the syllogism. Find out what you can about this mode of reasoning. To what extent does it correspond to the ways in which we actually make judgments?

C 4 Can a syllogism add to our knowledge—in any sense?

C 5 What is meant by induction? What part does it play in Scientific Method?

27

. . . gain Variety by Links or by Order

Back to the 'normal' English sentence:

The burly /	centre forward /	booted /	the ball /	into the net
adjective	*subject*	*verb*	*complement*	*adverb*

Without using any of the words from the given sentence (except 'the') make up three others after the same pattern—not all about sport, please.

But, of course, a piece of writing which consisted of sentences of exactly the same type would be very dull. No one would want to read a story which began:

(1) The burly centre forward booted the ball into the top of the net.

(2) The sprightly goalkeeper punched it to one side.

(3) The alert inside right slammed it out of the goalie's reach.

These three sentences considered together are monotonous and, although they are all made to the same pattern, they do not fit together either. The adjectives (pick them out) and the verbs, which give a clear picture of the action (pick them out too) give a little variety, but, as a piece of English, the three sentences together are bad.

One way in which the monotony of the three sentences can be lessened is to make them hang together by using linking words or phrases called *conjunctions* (compare 'junction'). The commonest and most useful of these words is 'and', but to use it on every occasion would be worse than useless. For example, if sentence (1) is joined to sentence (2) with 'and', it sounds as if the goalkeeper is helping the centre forward, whereas he is doing just the opposite.

Which word can be used instead?

There is contrast once more between sentences (2) and (3)—if you have already used 'but' between sentence (1) and (2) which word can be used for this second link?

Another way we could tackle the same sentence sequence would be to emphasise the order of events. The most obvious way of doing so would be to use the word 'then'—really an adverb taken out of order. Unfortunately 'then' comes into our minds all too easily. Which other words mean much the same as 'then'?

Make a list of words and phrases which can be used instead of 'and', 'then' and 'but', under the headings:

CONJUNCTIONS		ADVERBS
helping or supporting	*contrasting*	*time*
and	but	then
also	on the other hand	next

The list has been started for you.

A further way of getting variety is to vary the word-order. Everybody changes word-order naturally and easily in conversation. In written work, though, this varied effect is not so simply come by. Thus, it is good policy to write a thing as you would say it; say it to the person who is going to read your work, I mean: not (unless you *are* writing to a friend) as you would tell the story to your closest pal.

Taking this sentence as the basic one (normal order, but with no complement):

He / went / down the hill.
subject verb adverb

consider some possible changes of order and emphasis:

(a) Did / he / go / down the hill?
verb subject adverb

(b) Down the hill / he / went.
adverb subject verb

Or even,

(c) Went / down the hill, / he / did.
verb adverb subject verb

Which of these sentences puts most emphasis on the subject? Which stresses the action more than anything else? Which draws attention to the direction?

General rule: the beginning of the sentence has the most impact on the reader, so the part you wish to stress should come first.

Here are three commands—remember a command can be polite:

Get up, you.
Up you get.
You get up.

Which of them seems (a) most helpful, (b) to be a retort by someone who was himself being asked to get up, (c) most urgent, because most emphasis is put on the action—the person addressed being added only as an afterthought? Which word in the sentences is not strictly necessary?

A 1 What is the normal sentence-order?

A 2 Why depart from it?

A 3 What do you understand by 'monotony'?

A 4 What are two ways of avoiding monotony?

A 5 What are the commonest circumstances in which we do not begin the sentence with its subject?

A 6 Why are commands normally such short sentences?

A 7 What would you say was the difference between a command and a question?

A 8 Why should the word or phrase that comes first in the sentence have most impact? What is meant?

B 1 Mary and John and William and Jane and Jean were inseparable and they were quite different characters. Mary and Jane were tidy and John and William and Jean were all on the scruffy side, and Jean was a girl. John and William put pencils behind their ears and put chewing gum in their handkerchiefs and had inky imitations of tattoo marks on their wrists and Jean was not so bad and she got into trouble most. She was a girl.

Make sense of the above paragraph by replacing six of the 'ands' with commas; missing out one 'put'; inserting 'yet' after an 'and'; changing another 'and' to 'but', using 'though' and 'although' for others, and inserting 'because', instead of a full stop.

B 2 Write a short story in which each of the commands: 'Get up you'; 'You get up'; and 'Up you get' are used. The words spoken must be in character.

C 1 This is much the way some small children tell a story:

Then I got up, then I had my breakfast, then I read the cereal packet. Then I went to the Saturday matinée and there were these two comedians. Then the fat one says to the thin one....

(a) Continue the story as the child might tell it—using 'then' in practically every line.

(b) Turn the story into a business-like account, replacing 'then' by other words as far as possible.

(c) Compare the number of word-repetitions in accounts (a) and (b). Which classes of words are used most often in account (b)?

28

Function and Content

A Voyage to Laputa

'We next went to a school of languages, where three professors sat in consultation upon improving that of their own country.

'The first project was to shorten discourse by cutting polysyllables into one, and leaving out verbs and participles, because in reality all things imaginable are but nouns.

'The other project was a scheme for entirely abolishing all words whatsoever; and this was urged as a great advantage in point of health as well as brevity. For it is plain that every word we speak is in some degree a diminution of our lungs by corrosion, and consequently contributes to the shortening of our lives. An expedient was therefore offered, that since words are only names for *things*, it would be more convenient for all men to carry about them such things as were necessary to express the particular business they are to discourse on... which has only this inconvenience attending it, that if a man's business be very great, and of various kinds, he must be obliged in proportion to carry a greater bundle of things upon his back, unless he can afford one or two strong servants to attend him. I have often beheld two of these sages almost sinking under the weight of their packs, like pedlars among us; who, when they met in the streets, would lay down their loads, open their sacks, and hold conversation for an hour together; then put up their implements, help each other to resume their burdens, and take their leave....

'The day of our departure being come, I took leave of his Highness the Governor of Glubbdubdrib, and returned with my two companions to Maldonada, where after a fortnight's waiting, a ship was ready to sail for Luggnag....

'The dispatch came from court about the time we expected. It contained a warrant for conducting me and my retinue to Traldragdubh or Trildrogdrib, for it is pronounced both ways as near as I can remember.... When I had crept within four yards of the throne, I raised myself gently upon my knees, and then striking my forehead seven times upon the ground, I pronounced the following words, as they had been taught me the night before, *Ickpling gloffthrobb squutserumm blhiop mlashnalt zwin tnodbalkguffh slhiophad gurdlubh asht*. This is the compliment established by the laws of the land for all persons admitted to the King's presence. It may be rendered into English thus: May your Celestial Majesty outlive the sun eleven moons and a half.'

From *Gulliver's Travels*, by Jonathan Swift

A 1 What seems to be the common basis of the two schemes for reforming language?

A 2 At which point does it become clear that Swift is not seriously advocating these schemes?

A 3 How does Swift bring out the absurdity of the proposals?

A 4 Swift seems to be making two points through Gulliver's speech to the King—what are they?

A 5 In general what do Swift's views on language appear to be?

B 1 'We next went to a school....' Let us keep the structure, but vary the sense:

Pronoun	*Adverb*	*Verb*	*Preposition*	*Article*	*Noun*
I	then	got	into	an	academy
you	afterwards	jumped	over	a	university
he	frequently	travelled	out of	the	building
she	usually	leapt	down		college
they	subsequently	ran	from		edifice
it	daintily	strolled	behind		nursery
we	helpfully	marched	beyond		block
Mary	cheerfully	arrived	at		playground

(a) Which columns might be extended indefinitely?

(b) Which column is plainly as complete as it can ever be?

(c) Which column is almost complete? Regardless of sense, add the other words of the same type.

(d) Which column, having completed one series, has (illegitimately) gone on to another?

(e) What have the columns mentioned in (b), (c) and (d) in common?

B 2 The columns which we are able to complete we can call 'closed' systems, those which could always be added to (by coined words, even) are classified as 'open'. When we talk of vocabulary we automatically think of words in the 'open' class: *content* or *lexical words* are the terms used. Conjunctions, prepositions, pronouns, and articles are known as *function* or *grammatical words*: they provide what one might call the scaffolding of language. It is they, incidentally, which help to make the ''Twas brillig' type of nonsense *sound* meaningful. Remove pronoun, conjunction and article and we are left with, 'was brillig slithy tove', which sounds like a truncated question. Omit the real word 'was'—what remains is not nonsense, but no-sense.

B 3 All the prepositions, conjunctions, pronouns and articles in the language could be written down in a few minutes by anyone with his wits about him. You should already have seen or written in B2 all the pronouns that can be the subject of a sentence, all the prepositions, all the articles. Write down now all the conjunctions you can think of. It will seem easier if you take Swift's second sentence and substitute other words for 'and'. You might try now one 'open' word—how many near-synonyms can you find for 'said' in ten minutes?

B 4 In the extract from *Gulliver's Travels* how many grammatical/function words does Swift use? What is the proportion of function words to content words? How frequent are the repetitions in the case of (a) grammatical/function words, (b) lexical/content words?

B 5 If we accepted the first 'proposal', but retained verbal nouns and did not abbreviate, we should cut Swift's writing down to a quarter of its present length. The first sentence would read: 'School languages professors consultation improving country.' To be sure, that statement makes as much sense as some headlines; this is because certain words still 'hang together', namely, 'school languages' and 'improving country'. But remove the characteristic endings for plural and participle and we get, 'school language improv country'. Even now there is some sort of meaning, though not the original one.

Cut down the next two sentences in similar stages. Try to decide to what extent sense depends here on (a) grammatical/function words, (b) word-endings, (c) word-order—all of which are the business of grammar: the content is a question of vocabulary.

C 1 All writers must use function words. Attempt to find 'favourite' *content words* from sample pages of modern writers. An interesting comparison might be obtained if the work were done with reference to P. G. Wodehouse and Ernest Hemingway. What do you notice about the *style* of your selected writers? Is it in accord with the toughness or flippancy etc. of the vocabulary?

C 2 In the eighteenth century, Swift and other prominent writers hoped for 'an Academy that by the best Authorities and Rules drawn from the Analogy of Languages should settle all Controversies between Grammar and Idiom'. Is it possible or desirable to 'fix' a language? Are there countries where this has been attempted—with how much success?

C 3 How would we set about checking whether a series of utterances was part of a language or just so much gibberish? Consider Gulliver's compliment to the King. Is there enough evidence here?

29

Pattern of Usage

William Legrand in Edgar Allan Poe's story, *The Gold-Bug*, has uncovered the strange series of characters given below:

53‡‡†305))6*;4826)4‡.)4‡);806*;48†8¶60))85;I‡(;:‡*8†83(88)5
*†;46(;88*96*?;8)*‡(;485);5*†2:*‡(;4956*2(5*–4)8¶8*;4069285);
)6†8)4‡‡;I(‡9;48081;8:8‡1;48†85;4)485†528806*8I(‡9;48;(88;4
(‡?34;48)4‡;I6I;:188;‡?;

Having established that the message is a cipher and, from external evidence, that the cipher is in English, Legrand gives an account of how he set about solving it.

'You observe there are no divisions between the words. Had there been divisions, the task would have been comparatively easy. In such case I should have commenced with a collation and analysis of the shorter words, and had a word of a single letter occurred, as is most likely ('a' or 'I', for example), I should have considered the solution as assured. But, there being no division, my first step was to ascertain the predominant letters, as well as the least frequent. Counting all, I constructed a table thus:

Of the characters	there are	Of the characters	there are	Of the characters	there are
8	33	6	11	:	4
;	26	†	8	3	4
4	19	I	8	?	3
‡	16	0	6	¶	2
)	16	9	5	–	1
*	13	2	5	.	1
5	12				

'Now, in English, the letter which most frequently occurs is "e". Afterwards, the succession runs thus: a o i d h n r s t u y c f g l m w b k p q x z. "e" predominates so remarkably, that an individual sentence of any length is rarely seen, in which it is not the prevailing character.

'Here, then, we have, in the very beginning, the groundwork for something more than a mere guess. The general use which may be made of the table is obvious—but in this particular cipher we shall only very partially require its aid. As our predominant character is 8 we will commence by assuming it as the "e" of the natural alphabet. To verify the supposition, let us observe if the 8 be seen often in couples—for "e" is doubled with great frequency in English—in such words, for example, as "meet", 'fleet", "speed", "seen", "been", "agree", &c. In the present instance we see it doubled no less than five times, although the cryptograph is brief.

'Let us assume 8, then, as "e". Now, of all *words* in the language, "the" is most usual; let us see, therefore, whether there are not repetitions of any three characters, in the same order of collocation, the last of them being 8. If we discover repetitions of such letters, so arranged they will most probably represent the word "the". Upon inspection, we find no less than seven such arrangements, the characters being ;48. We may, therefore, assume that ; represents "t", 4 represents "h", and 8 represents "e"—the last being now well confirmed. Thus a great step has been taken.

'But, having established a single word, we are enabled to establish a vastly important point; that is to say, several commencements and terminations of other words. Let us refer, for example, to the last instance but one in which the combination ;48 occurs—not far from the end of the cipher. We know that the ; immediately ensuing is the commencement of a word, and, of the six characters succeeding this "the", we are cognisant of no less than five. Let us set these characters down thus, by the letters we know them to represent, leaving a space for the unknown—

t eeth

'Here we are enabled at once to discard the "th", as forming no portion of the word commencing with the first "t"; since, by the experiment of the entire alphabet for a word adapted to the vacancy, we perceive that no word can be formed of which this "th" can be a part. We are thus narrowed into

t ee,

and, going through the alphabet, if necessary, as before, we arrive at the word "tree", as the sole possible reading. We thus gain another letter, "r", represented by (, with the words "the tree" in juxtaposition.

'Looking beyond these words, for a short distance, we again see the combination ;48, and employ it by way of *termination* to what immediately precedes. We have thus this arrangement:

the tree ;4(‡?34 the,

or, substituting the natural letters, where known, it reads thus:

the tree thr‡?3h the

'Now, if, in place of the unknown characters, we leave blank spaces, or substitute dots, we read thus:

the tree thr...h the,

when the word "through" makes itself evident at once. But this discovery gives us three new letters, o, u and g, represented by ‡? and 3.

'Looking now, narrowly, through the cipher for combinations of known characters, we find, not very far from the beginning this arrangement, 83(88, or egree, which plainly is the conclusion of the word "degree", and gives us another letter, d, represented by †.

'Four letters beyond the word "degree", we perceive the combination:

;48(;88

'Translating the known characters, and representing the unknown by dots, we read thus:

th.rtee.,

an arrangement immediately suggestive of the word "thirteen", and again furnishing us with two new characters "i" and "n", represented by 6 and *.

'Referring, now, to the beginning of the cryptograph, we find the combination,

53‡‡†.

'Translating, as before, we obtain

.good,

which assures us that the first letter is A, and that the first two words are "A good".'

From *The Gold-Bug*, by E. A. Poe

A 1 What do you suppose made Legrand so certain that the numerals and signs were a cryptogram?

A 2 What is the purpose of the table of characters?

A 3 Give the stages by which it becomes certain that 8 is 'e'.

A 4 Why is the identification of any whole word of particular importance?

A 5 What proportion of Legrand's work depends on (a) knowledge of language, (b) orderly procedure, (c) trial and error? Give details.

B 1 If there had been no 'external' evidence (a signature, in point of fact) of the language in which the message was written, what sort of 'internal' considerations might help to determine whether the language was English or some other?

B 2 Set down the characters which Legrand has identified together with their equivalents. Complete the solution of the cryptogram and extract the message. Check your answer (it will still sound fairly cryptic) by reading the story.

C Until recently, the sort of script shown on page 118 could not be read. See *The Decipherment of Linear B*, by

J. Chadwick—a 'Pelican' book which has the interest of a cryptogram and historical significance too.

If you are more concerned with literature, read *The Shakespeare Ciphers Examined*, by W. F. and E. Friedman, two American cipher experts. Using the methods of Baconians and others, the authors extract from the plays of Shakespeare such messages as, 'I Francis Bacon, wrote these plays. No kidding.' However, Bacon did invent at least one cipher of his own (see a translation of the Latin version of the *Advancement of Learning*) and Mr and Mrs Friedman have, apparently, embodied a hidden message of their own in this very amusing book—so there is something to search for.

30

A Persuasive Tongue

If you did not find out earlier what a syllogism was, here is a suspect example:

A loaf of bread is better than nothing.
Nothing is better than the Kingdom of Heaven.
∴ A loaf of bread is better than the Kingdom of Heaven.

An outrageous conclusion, from an argument which imitates the lines of:

Brown's writing is better than Smith's
Smith's writing is better than Jones'
∴ Brown's writing is better than Jones'.

But, while Smith's writing is a *thing* that can play its part in a comparison, 'nothing' is not. So no conclusion is possible. Anyway the second line of the argument is deliberately arranged so as to confuse—if it is rephrased as: 'the Kingdom of Heaven is above all things', the opposite conclusion to the one arrived at would be the only possible one. Rearrange the argument so as to produce a conclusion about the superiority of Heaven to a loaf of bread.

The first argument is a frivolous one, but it does show how we can almost be persuaded by the form of words into accepting something ridiculous. In the sentence: 'A swan is white', we certainly refer to the swan. In the sentence: 'The dragon is non-existent', we seem to be following the same pattern. Do we refer to the dragon as a thing? No, as an idea, merely; for 'existence' is not something that a thing can *have*, in the way that a hen has feathers and a cat fur. 'Existence', as philosophers say, 'is not an attribute'.

On a less philosophic plane, it is worth thinking about the effect words have: why we or other people decide to depart from the 'normal' sentence, or choose one particular word or phrase out of all those possible. (Our language, because of its Germanic origin and the country's close ties with France, is, as I hope has been shown, rich in equivalent words or synonyms.)

Advertisers—and politicians—have made such a study and it is as well for us to try to match what they are selling us or telling us with the possible reality.

For instance when an advertiser says: you DESERVE a something or other—stating the name of his product, what is he implying about: (1) his product? (2) you? As he cannot know you, yet wishes to get you to buy, what may the advertiser be said to be doing?

Suppose a large empty house is up for sale. Various people from their different viewpoints have described it as:

large, huge, barn-like, roomy, spacious, vast,
enormous, commodious, sizeable.

Which of these adjectives would the estate agent be likely to use? Which would be least likely to sell the house? Which of the words might be described as 'neutral'?

To a large extent, it is the 'associations' words have (what comes into our heads when the word is mentioned) that determine their effect. 'Huge' makes us think of giants, ogres, castles—ugly bigness; 'barn-like', of draughts, owls, straw, rough beams—and so on.

Associations can be changed for good or ill by advertising and other publicity. When Robert Burns wrote about 'woodbine twine' he expected the word 'woodbine' to make us think of hedgerows, open air and sweet scent. What association has the word 'woodbine' for you today? What did Burns mean? Why the change?

Explain how the association of 'zebra' has changed over the last few years.

Synonyms can have very different effects—again, partly a question of association: 'The well-known play-boy, Aloysius Spinebeak, spends his leisure hours collecting bric-à-brac....' The beginning of this imaginary report could, without actually lying be recast maliciously, so as to read: 'This notorious wastrel idles his life away amassing rubbishy bits and pieces....'

A 1 In your own words, what is wrong with the reasoning in the loaf–Heaven argument?

A 2 Why is it suggested that we should keep a close watch on how words are used?

A 3 Explain what is meant by 'word-association'.

A 4 What is a synonym? Why is our language unusually rich in synonyms?

A 5 What is implied by the phrase 'not actually lying'?

B 1 Rewrite these two descriptions, replacing the words in italics by near-synonyms, so as to give an unfavourable impression of Mary and her house and of Bill Teak's tactics:

Provocative Mary Whalebone, a *dashing* rider in her youth, has recently *acquired* a *bijou residence* in one of the *less-fashionable* areas of London. The *kitchen* garden *needs some attention*, but the *period cottage* was *excellent value* and is a *veritable sun trap*.

The *robust* tackling of *forthright* Bill Teak, one of soccer's *veterans* has *been commented on by* visiting wingers. Although *not the speediest* man in the game, Bill and his policy of '*first-time*' *play* have often *reduced the odds against his own team*.

B 2 Here are some 'neutral' words and phrases. Find expressions which mean the same, but are (a) approving, (b) disapproving:

thin, over-weight, bright (clothes), up to date, old-style, fair (hair), tall, short, careful (with money), a revolutionary, quiet (of a person), quiet (of a place), food-loving, firm (character), warm (weather).

Set them out as shown; the first one has been done for you:

NEUTRAL	APPROVING	DISAPPROVING
thin	slim, slender	skinny

C 1 Find out the cost of advertising on ITV at a peak-viewing time.

C 2 Are advertisements on television more dependent on sound or on visual effects, or do they vary? Give details.

C 3 Find examples of television adverts which (a) have a dream-like sensuous appeal, (b) are full of 'impact' words and are very masculine, (c) make use of humour even to the extent of poking fun at the product, (d) show people at home using the goods, (e) mix this documentary effect with fantasy, (f) have a 'second-voice' which backs up the point of the jingle or playlet. (These categories are not, of course, mutually exclusive.)

C 4 Find out what you can about the *Daily Herald*. Why has this newspaper, which once had the largest circulation in Britain, had to make way for *The Sun*?

C 5 To end this section, here are some samples of advertising matter which appeared in British newspapers and magazines during 1964. The extracts have been chosen for their verbal interest rather than for any 'stunt' quality. 'Prestige' advertising has not been included here, though it is well worth investigating.

(i) Fun is catching with the getaway people. Laughing, smiling, great-to-be-alive enjoyment. Getaway people laugh easily, but take their choice of petrols seriously.

(ii) Wine's a Winner with Sunday dinner.

(iii) Gay up your day the pocket-size way—with music to match the mood. Pocket-size portables are perfect for radio pleasure.

(iv) It's the orangemostest drink in the world.

(v) There's something familiar about this new cigarette.
And if you're half the smoker we think you are,
You'll recognise it first puff.
That's right it's the taste.
It comes through nice and even, redolent of really good tobacco. Most enjoyable and quite unmistakable.

(vi) It's one of those things—like first night tickets
Or a personal welcome from the headwaiter—
that tells people just how much you expect out of life.

(vii) Amplify your voice—Make them listen.

(viii) A tiny power cell energizes a transistorized circuit. The circuit sends pulses of current through two coils, consisting of 600 feet of wire 1/5th the thickness of a human hair. The magnetic field thus created attracts and repels the tuning fork at the precise rate of 360 vibrations a second—to give the electronic watch its remarkable guaranteed 99·9977% accuracy.

(ix) Every time your skin dries a little it dies a little.

(x) A purry, furry, jerkin—feline! fascinating! Wickedly attractive with genuine jungle markings. Sleek cut, lined with a silky, slinky flash of sizzling scarlet.

(xi) Cider for the sun set. Who are the sun set? People like you who chase the sun, who give refreshment its proper consideration and thirst its due reward. And the sun set's favourite sundowner is the golden taste of cider.

(xii) It's New—Designed for 'dreamagers'—double nylon briefs 'n' bra with fabulous frothy frilling. A captivating dreamy sleepcoat completes the picture.

(xiii) So who wants 9·7% more acceleration? You do. In traffic, acceleration is safety....

(xiv) Your children, daily growing more endearing; more enquiring. Daily developing into the sort of people you want them to be. Nothing can be more important to a mother than to watch them grow from utterly defenceless babyhood to happy healthy childhood. Nothing can be more inspiring than to know that what they are and will be depends on you more than anyone else in the world. That's why you give them, etc.

(xv) Always put a pinta on the table.

(xvi) Relax on golden beaches. Sit down to glorious meals. Meet laughing friendly people. Drive on traffic-free roads. Enjoy superb hotels. Golf, fish. It all adds up to a happy holiday with no language or currency worries.

(xvii) At last—your own idea of what a superb modern hotel should be like. Every room has private bath and shower ...full central heating...food, service, and surroundings are top flight throughout, from the Lucullian menus of the Embassy Room to the startling view from the Martello Grill, high over the City. Relax in the quiet luxurious cocktail lounge...above all, enjoy the skilled service and courtesy that marks our Hotels the world over.

(xviii) Goodwill train tour of People's China.

(xix) Ideal boarders. Home cooking. Interior springs. H & C lounge, TV. Terms moderate—Write brochure.

(xx) Deutsche Buecher Gesucht.

After making a close study of the style of each advertisement and taking into account the implied 'image' of the person addressed, place the advertisements in sequence according to (a) probable effectiveness, (b) subtlety (or lack of it). Justify the order you decide upon.

31

Language and Thought

'Last week, as I stood shaving, the drop fell. I, standing with my razor in my hand, became suddenly aware of the merely habitual nature of my action (this is the drop forming) and congratulated my hands, ironically, for keeping at it. Shave, shave, shave, I said. Go on shaving. The drop fell. All through the day's work, at intervals, my mind went to an empty place, saying, "What is lost? What is over?" And "Over and done with," I muttered, "over and done with," solacing myself with words. People noticed the vacuity of my face and the aimlessness of my conversation. The last words of my sentence tailed away. And as I buttoned on my coat to go home I said more dramatically, "I have lost my youth."

'It is curious how, at every crisis, some phrase which does not fit insists upon coming to the rescue—the penalty of living in an old civilisation with a notebook.'

From *The Waves*, by Virginia Woolf

A 1 The 'drop' probably came from a tap. What, though, does it seem to represent?

A 2 Suggest two reasons for the repetition of words and phrases in the extract.

A 3 What evidence is there that the character concerned is highly conscious of the workings of his own mind, but has a sense of humour which stops him being too self-important?

A 4 Attempt an explanation of the last sentence of the extract.

A 5 The first phrase which occurs to him and does not quite match the experience is 'over and done with'; what is the second?

B 1 What is likely to be the effect on a thought of finding a phrase which *almost* expresses it?

B 2 Write down half a dozen advertising or political slogans. Which seem to be backed by public opinion?—or is this 'backing' partly the result of similar campaigns? 'Public opinion' is something of a slogan itself, suggesting that what the ordinary man (and who is he?) thinks, matters; that the population has a unity, a personality.

B 3 By what means have writers tried to convey the thoughts of their characters, apart from the dodge of treating thought as if it were conversation? Do you yourself think in words, images—or is it a mystery to you?

Can you think of two things at once? Try to read from one book at the same time as you are writing down something quite different. Note the mistakes that you make.

C 1 People seem to believe that men and women of different nationalities tend to have different thought processes. To what extent does it seem plausible to attribute such differences to the structure of one's native tongue?

C 2 Are you yourself often stuck for words or do words (not necessarily the right ones) come all too readily? If you are called on to speak do you try to consider how you would write what you want to say—or, do you rather, if you have to write, think in terms of speech?

32

Off the Record

A little while ago, I was going over some work with a student and, at one point, criticised her use of the word 'vague' when she should have put 'vagueness'. Worse, she seemed to have spelt it 'vauge'. The student explained that the word was actually 'range', which made good sense.

My recollection of the next part of the conversation was that it had gone like this:

(a) 'Oh, it's your writing. I take it all back. I won't cross out my comment, but you can remember what I said. I was a bit puzzled by what you meant, for that "r" isn't very clear, though I agree you can form one in that way. Were you taught the italic style at school?'—No. 'This sort of misunderstanding is common when you're reading *children's* handwriting. Here's an example which I remember. A girl wrote this: —(). That says "mum". Do you see?'—Oh, yes. 'An italic hand can be harder to make out than any other form of writing.'

A tape-recorder, which had been 'on' in the room for some time, preserved what I actually said. It is given below:

(b) 'Oh ah it it's your writing oh I take it all back well you can remember that I've taken it back huhu—sn—ah yeah I   a bit puzzle' but this this isn't a very good "r" I agree that you can do it like that um   ah um this italic writing mmmm w-were you um taught um this er italic style or is it something you've um picked up   ah you get it with children you know this italic style and it tends to come out like this if it's badly done ah I remember this ah from a girl   that says "mum"   can you see   y'know it-it's worse than any sort of writing to read if it goes a bit astray   Um.'

A 1 Trace the changes of direction in the thought of passage (b).

A 2 How have they been smoothed out in passage (a)?

A 3 Suggest why passage (b) when spoken seemed more intelligible than it looks.

A 4 Pick out some of the structural characteristics of speech from passage (b).

A 5 What are the chief difficulties which face a writer when he attempts to put speech on paper?

B Here are some examples of 'speech on paper'—all stylised in varying degrees. To make comparisons easier, each concerns a man and a woman, usually in the role of lovers.

(a) *Stychomythia—rare in English*

Anne: I would I knew thy heart.
Gloucester: 'Tis figured in my tongue.
Anne: I fear me both are false.
Gloucester: Then never man was true.
Anne: Well, well, put up your sword.
Gloucester: Say then, my peace is made.
Anne: That shalt thou know hereafter.
Gloucester: But shall I live in hope?
Anne: All men, I hope, live so.
Gloucester: Vouchsafe to wear this ring.
Anne: To take is not to give.

From *Richard III*, by William Shakespeare

(b) *Formal disputation*

'Marriage is evidently the dictate of nature; men and women are made to be companions of each other; and therefore I cannot be persuaded but that marriage is one of the means of happiness.'

'I know not,' said the princess, 'whether marriage be more than one of the innumerable modes of human misery. When I see and reckon the various forms of connubial infelicity, the unexpected causes of lasting discord, the diversities of temper, the oppositions of opinion, the rude collisions of contrary desire where both are urged by violent impulses, the obstinate contests of disagreeing virtues where both are supported by consciousness of good intention, I am sometimes disposed to think with the severer casuists of most nations, that marriage is rather permitted than approved, and that none, but by the instigation of a passion too much indulged, entangle themselves with indissoluble compacts.'

'You seem to forget,' replied Rasselas, 'that you have, even now, represented celibacy as less happy than marriage. Both conditions may be bad, but they cannot both be worst.'

From *Rasselas*, by Dr Johnson

(c) *'Involved' realism*

'...it is the law of creation. One is committed. One must commit oneself to a conjunction with the other—for ever. But it is not selfless—it is a maintaining of the self in mystic balance and integrity—like a star balanced with another star.'

'I don't trust you when you drag in the stars,' she said. 'If you were quite true, it wouldn't be necessary to be so far-fetched.'

'Don't trust me then,' he said, angry. 'It is enough that I hurt myself.'

'And that is where you make another mistake,' she replied. 'You *don't* trust yourself. You don't fully believe yourself what you are saying. You don't really want this conjunction, otherwise you wouldn't talk so much about it, you'd get it.'

He was suspended for a moment, arrested.

'How?' he said.

'By just loving,' she retorted in defiance.

From *Women In Love*, by D. H. Lawrence

(d) *Two styles*

'Oh, Bertie, you remind me of Rudel.'

The name was new to me.

'Rudel?'

'The Seigneur Geoffrey Rudel, Prince of Blaye-en-Saintonge.'

I shook my head.

'Never met him, I'm afraid. Pal of yours?'

'He lived in the Middle Ages. He was a great poet. And he fell in love with the wife of the Lord of Tripoli.'

I stirred uneasily. I hoped she was going to keep it clean.

'For years he loved her, and at last he could resist no longer. He took ship to Tripoli, and his servants carried him ashore.'

'Not feeling so good?' I said, groping. 'Rough crossing?'

'He was dying. Of love.'

'Oh ah.'

From *The Code of the Woosters*, by P. G. Wodehouse

(e) *The interior monologue*

...the sun shines for you he said the day we were lying among the rhododendrons on Howth head in the grey tweed suit and his straw hat the day I got him to propose to me yes first I gave him the bit of seedcake out of my mouth and it was leap-year like now yes 16 years ago my God after that long kiss I near lost my breath yes he said I was a flower of the mountain yes so we are flowers all a womans body yes that was one true thing he said in his life and the sun shines for you today yes that was why I liked him because I saw he understood or felt what a woman is and I knew I could always get round him and I gave him all the pleasure I could leading him on till he asked me to say yes and I wouldnt answer first only only looked out over the sea and the sky....

From *Ulysses*, by James Joyce

(f) *Realism*

At our stop I stand back and wait for her to get off. I try to give her a smile but I can feel it doesn't amount to much.

'Thanks for saving me life.'

'Oh, that's all right.'

We start off up the lane to the Works.

'I'll see you get it back as soon as I get some money.'

'Oh, don't worry about it.'

'I never picked it up off the dressing-table. I had sixpence in my overcoat pocket so I never noticed till I was on the other bus, y' see. One of the lads'll help me out.'

From *A Kind Of Loving*, by Stan Barstow

(g) *Inside and out*

'Who do you love?' said Liz.

'Thee, lass,' I said, finding refuge in the Duxbury dialect.

'Yes, it sounds like it, doesn't it?'

'Ah do, lass.'

'Say it properly, then.'

'I do, Liz, I do,' I said soberly, and wondering if I meant it.

From *Billy Liar*, by Keith Waterhouse

B 1 Consider passage (a) in relation to passage (b), passage (b) in relation to (c), and (c) in relation to (d). Say in each case what their form and manner have in common and how they differ.

B 2 Compare passages (d), (e), (f) and (g) with the characteristics of speech as revealed by a tape-recorder. What are your conclusions?

B 3 Compare passages (e) and (g) and a tape-recorded conversation, showing what they imply about our thought processes.

B 4 How is the effect of realism achieved in passage (f)? The author could justifiably claim that this exchange might in real life be almost as fluent—on what grounds?

B 5 Which passage do you consider provides the most successful blend of nature and art? Remember, even the assumption that people *take turns* in speaking is a literary convention. Where is artifice apparent in the handling of passage (e)?

C 1 Attempt to take down a conversation verbatim, including all the ums and ers and gaps. Use a tape-recorder if possible and insert no punctuation other than space for noticeable pauses. How much of what was said was not strictly necessary? Does it seem that there were certain words or phrases which became 'fixed' in the speaker's mind, and so recurred? What other characteristics of 'real' speech do you note?

C 2 Read the verses given below. Which characteristics of 'real' spoken English do you find here? How does it differ from poems which owe little to the folk tradition?

Kyng Adland hath a daughter, brother,
  Men call her bright and sheene;
If I were kyng here in your stead,
  That ladye shold be my queene.

Saies, Reade, reade me, deare brother,
  Throughout merry England,
Where we might find a messenger
  Betwixt us towe to sende.

Saies, You shal ryde yourselfe, brother,
  Ile beare you companye;
Many a man is throughe fals messengers deceived,
  And I feare lest soe shold wee.

Thus the renisht them to ryde
  Of twoe good renisht steeds,
And when the came to kyng Adlands halle,
  Of redd gold shone their weeds.

And when the came to kyng Adlands halle
  Before the goodlye gate,
There they found good kyng Adland
  Rearing himselfe theratt.

Now Christ thee save, good kyng Adland,
  Now Christ you save and see.
Sayd, You be welcome, kyng Estmere,
  Right hartilye to mee.

You have a daughter, said Adler younge,
Men call her bright and sheene,
My brother wold marrye her to his wiffe,
Of Englande to be queene.

From *The Ballad of King Estmere*

How do you account for, on the one hand, the 'speech' elements; on the other, rhyme and rhythm? Consider the ballad form in relation to the times which produced it.

C 3 Compare the first two stanzas of this extract with the opening of *Sir Patrick Spens*. Compare the ballad *Edward* with *Lord Randall* and *The Twa Corbies*. What similarities do you notice? In much the same way that dialects develop from a parent language, variations (which become separate songs or poems) develop from a few basic tales. That is, if they are passed on (by word of mouth) in the traditional way and not 'fixed' by print. Explore either the medieval ballad or the American 'blues' and consider the suggestion (made, actually, about the blues) that all the existing material may stem from *one* basic song.

33

Vital Statistics

'They have been severely punished', says the headmaster, wishing to imply that justice is impersonal. In spite of the missed opportunity for additional emphasis, he would hardly use the construction 'Punishment has been given them'.

This second arrangement—the original form of the passive voice in English—*is* sometimes used. For instance, the statement 'She was given a handbag as a leaving present' might be re-expressed as 'A handbag was given to her as a leaving present'. However, the first construction is the more common, for it leaves undisturbed our most usual sequence of ideas: person + verb + further information. It sounds less than natural in English to use an inanimate object or an abstract quality as the subject if there is a pronoun in the sentence (never mind its grammatical form!) which can be called on to do the job.

In none of the examples used so far is the grammatical subject the instigator of any sort of action, but the word-order in each case follows the familiar SVC or SVA pattern. This sequence is less familiar in some languages of the Indo-European group—while in Chinese it is, apparently, more or less inevitable. In German or in Old English only about 30 per cent of sentences in which the elements SVC occur deal with them in that order. In modern French, the proportion is about 70 per cent.

The commonest instance of altered word-order in English is when one asks a question. The simplest way of asking a question is to invert the order of subject and verb. Thus the statement 'you are' becomes the question 'are you?'. However, the construction is more widely used in French and German; for English takes unkindly to a VS arrangement. We cannot turn, for example, 'He eats his breakfast' into 'Eats he his breakfast?' We say, rather, 'Does your little boy eat his breakfast?' or 'Is he eating his breakfast?'—or even, 'He eats his breakfast, does he?' Sometimes, in conversation, we simply leave the sentence as it is: 'He eats his breakfast?' In this case, though, we probably add some meaningless word such as 'then' to the end of the question.

A 1 Explain why 'They have been severely punished' might imply that justice was impersonal. Suggest other reasons and occasions for preferring the passive voice to the active.

A 2 Suggest contexts for each of the sentences about the handbag presentation. If possible, the version in question should be obviously more suitable than the alternative.

A 3 Account in general terms for the statistics about SVC sentences in French and German. What stylistic disadvantage would an 'inevitable' SVC sentence-order have?

A 4 In English, which verbs besides 'to be' can 'ask questions' by an inversion of subject and verb?

A 5 'He eats his breakfast, then?' In such a sentence, 'then', unless it were given unexpected stress, has nothing to do with time. Why, then, should the addition of such a word help that (and this) sentence along?

B 1 Pick a page or two of modern prose at random. What proportion of all sentences used there contains the elements SVC? Of those that do, how many are strictly S–V–C in order? List the variations in order of frequency.

B 2 Note also the frequency of, on the one hand: adjectives before the noun; comparative forms of adjectives using the suffixes -er, -est; negative formed by using 'no' as an adjective; vowel change to form plural; possession indicated by apostrophe. On the other: adjectives after the noun (not clauses); comparative forms of adjectives, using 'more', 'most'; negative formed by insertion of 'not'; plurals indicated by -s; possession indicated by 'of'.

What conclusions do you come to regarding the German/French elements in English syntax?

B 3 The verbs which you should have written down in A5 are (even in the case of 'to be' and 'to have') used chiefly as auxiliaries. Choose a passage of conversation from a modern writer and note the use of 'is', 'have', 'do', 'shall', 'can', 'might', 'would' and related forms. What proportion of their usage consists in forming compound verbs? Even when they stand alone, how often is the compound form implied? What do you see as (a) the grammatical function, (b) the stylistic contribution of these verbs?

C Add to the passages you have used so far (include those reproduced in the previous chapter if you wish): an extract from a school science book; a newspaper report on an exhibition of paintings; a guide book; a poem by Ogden Nash; Instructions for filling in a Tax Return.

Analyse each passage on a basis of literal and non-literal usage (including irony and conscious and unconscious use of metaphor). At what point does metaphor become cliché? The expression 'dead metaphor' is probably a cliché, and the term cliché (look it up) a dead metaphor. What is the difference between the two?

From the results of your analysis, what would be the loss statistically if we had to use words in a strictly literal way? Might there be a corresponding gain in clarity and precision—or would communication become almost an impossibility?

Books For Reference and Further Reading

Publication dates given for dictionaries are those of the latest significant revision, if any.

The Shorter Oxford English Dictionary, ed. Little, W., Fowler, H. W. and Coulson, J. Oxford University Press, 1956.

The Concise Oxford Dictionary, ed. Fowler, H. W. and Fowler, F. G. Oxford University Press, 1951.

A Dictionary of Modern English Usage, ed. Fowler, H. W. (revised Gowers, E.). Oxford University Press, 1965.

Roget's Thesaurus of English Words and Phrases, ed. Dutch, R. A. Longmans, 1962.

The Concise Oxford Dictionary of English Place-Names, ed. Ekwall, E. Oxford University Press, 1946.

The Oxford Dictionary of English Christian Names, ed. Withycombe, E. G. Oxford University Press, 1947.

A Dictionary of British Surnames, ed. Reaney, P. H. Routledge, 1958.

Dictionary of French and English, ed. Bellows, J. and Bellows, W. Longmans, 1948.

Pocket Dictionary of the German and English Languages, ed. Wichmann, K. Routledge, 1946.

* * * * *

Bradley, H., *The Making of English*. Macmillan, 1904.

Brook, G. L., *English Dialects*. Deutsch, 1963.

Fries, C. C., *The Structure of English*. Longmans, 1957.

Gowers, Sir E., *The Complete Plain Words*. (H.M.S.O., 1948.) Penguin, 1962.

Graves, R. and Hodges, A., *Reader Over Your Shoulder.* (Cape, 1947.) Mayflower, 1962.

Jespersen, O., *The Growth and Structure of the English Language.* Blackwell, 1905.

Levett, J., *The Spell of Words.* Darwen Finlayson, 1959.

Opie, I. and Opie, P., *The Lore and Language of Schoolchildren.* Oxford University Press, 1959.

Osmond, E., *From Drumbeat to Tickertape.* Hutchinson, 1960.

Partridge, E., *Usage and Abusage.* (Hamish Hamilton, 1957.) Penguin, 1963.

Potter, S., *Our Language.* Penguin, 1950.

Potter, S., *Modern Linguistics.* Deutsch, 1957.

Quirk, R., *The Use of English.* Longmans, 1962.

Rundle, S., *Language as a Social and Political Factor in Europe.* Faber, 1946.

Russell, B., *The Problems of Philosophy.* Oxford University Press, 1912.

Thompson, T., *The ABC of Our Alphabet.* Studio, 1942.